10 Colorado CMAS Grade 6 Math Practice Tests

The Ultimate Test Prep Collection with Answer Explanations

Dr. A. Nazari

10 Practice Tests

♛ The Grand Championship Collection ♛

Welcome, future Math Champion!

*You hold the **ultimate collection** —*
ten full-length practice tests designed to take you
*from first attempt to **complete mastery**.*

🏅 *Conquer every Grade 6 topic*

🏅 *Build unshakeable confidence*

🏅 *Rise from Bronze to Gold to Champion*

🏅 *Arrive at test day fully prepared*

The championship begins now.

★ ★ ★ ★ ★

> **66** *Ten tests may seem like a marathon, but champions are made one step at a time. Trust the process!* **99**

The Champion's Path

Bronze Round (Tests 1–3)

Your warm-up matches. Take these **untimed** to learn the format and set your baseline. Read the answer explanations after each test — this is where you build your foundation.

Silver Round (Tests 4–6)

Set a timer for **75 minutes**. Focus on the topics that tripped you up in Bronze. Practice showing your work on every problem. Your accuracy should be climbing.

Gold Round (Tests 7–9)

Full timed conditions (**60 minutes**). Simulate the real exam environment. Review only the questions you missed — targeted practice is the key to gold.

Championship Final (Test 10)

Your final match. Full exam conditions — timed, quiet, no breaks. This is your victory lap. Show yourself how far you've come!

Your Championship Kit

- **10 Full-Length Practice Tests** — every Grade 6 topic
- **Formula Reference Sheet**
- **Complete Answer Key** with explanations
- **Championship Scoreboard** to track your rise

Champion's Tip: Space your tests 2–3 days apart. Use the days in between for targeted review. By Test 10, you'll be amazed at your transformation.

👑 The Champion's Playbook 👑

I **Read every question twice.** *The first read tells you the topic. The second tells you exactly what to solve for. Champions never skim.*

II **Mark the clues.** *Circle key numbers, underline the question, and cross out information that's just there to distract you.*

III **Choose your strategy.** *Before touching pencil to paper, decide: Am I setting up a ratio? Solving an equation? Finding area? Name the approach.*

IV **Solve, then match.** *For multiple choice — work the problem on scratch paper first, then find your answer among the choices.*

V **Eliminate and conquer.** *Cross out obviously wrong answers. If you're left with two, you've already doubled your odds. Make an educated pick.*

VI **Estimate to verify.** *After solving, ask: "Is this answer reasonable?" A quick mental estimate catches most calculation errors.*

VII **Leave nothing blank.** *Even a well-reasoned guess is worth more than an empty space. Use partial work to support your answer.*

🕐 Timing Mastery

Tests 1–3: **Untimed** (build foundation) ▸ Tests 4–6: **75 min** (build speed) ▸ Tests 7–10: **60 min** (championship conditions)

⭐ Grade 6 Championship Topics

🏅 Ratios & Proportions   🏅 Integers & Rational Numbers   🏅 Expressions & Equations

🏅 Geometry & Measurement   🏅 Statistics & Data Analysis

*A true champion isn't someone who never makes mistakes — it's someone who learns from **every single one**. After each test, review your errors carefully. That's where the real growth happens.*

Find more at
ViewMath.com/CO-Grade6

 The Champion's Toolkit

Prepare your workspace before each championship round

🏆 Required Equipment

✏️	**Sharpened Pencils**	Two #2 pencils — champions always have a backup
🧽	**Quality Eraser**	A clean, soft eraser that won't smudge your work
📝	**Scratch Paper**	Blank paper for calculations, diagrams, and number lines
📏	**Ruler**	Essential for geometry and coordinate plane questions
⏱️	**Timer**	Begin using from the Silver Round onward
🔊	**Quiet Workspace**	A calm, well-lit area free from distractions

Permitted in Competition

- ✔ Pencil and eraser
- ✔ Scratch paper (provided)
- ✔ Ruler (if specified)
- ✔ Formula reference in this book

Not Permitted

- ✘ Calculators
- ✘ Electronic devices
- ✘ Textbooks or notes
- ✘ Outside help

- With 10 tests, space them **2–3 days apart**. This gives time to review mistakes and study between rounds.

- Let your child take Tests 1–3 untimed to build familiarity and establish a baseline.

- After each test, go through the Answer Key together. Focus on **understanding the reasoning**, not memorizing answers.

- Use the Championship Scoreboard to visualize long-term progress. Celebrate improvements at every tier!

- Pair with our **Grade 6 Math Study Guide** for topics that need sustained attention.

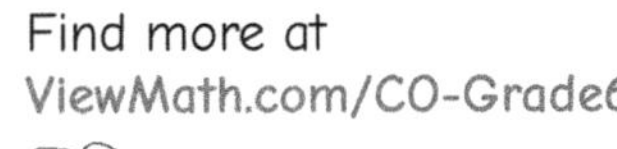
Find more at
ViewMath.com/CO-Grade6

Formula Reference Sheet

◢ Area Formulas

Rectangle	$A = l \times w$
Parallelogram	$A = b \times h$
Triangle	$A = \dfrac{1}{2} \times b \times h$
Trapezoid	$A = \dfrac{1}{2}(b_1 + b_2) \times h$

▣ Volume

Rectangular Prism $\quad V = l \times w \times h$

▣ Surface Area

Find the area of each face, then add them all up.

Rectangular Prism:

$SA = 2lw + 2lh + 2wh$

↓⅑ Order of Operations

P   Parentheses first

E   Exponents

M/D Multiply & Divide (left to right)

A/S Add & Subtract (left to right)

% Ratios & Percents

Ratio: $a : b$ or $\dfrac{a}{b}$

Unit rate: amount per 1 unit

Percent: a ratio out of 100

Part = Percent $\times$ Whole

⚖ Integers & Absolute Value

Integers:

$\ldots, -3, -2, -1, 0, 1, 2, 3, \ldots$

$|-5| = 5 \quad |5| = 5$

Absolute value = distance from 0

X¹ Expressions & Equations

Exponent: $3^4 = 3 \times 3 \times 3 \times 3 = 81$

Variable: a letter that stands for a number

Equation: two expressions joined by $=$

Inequality: uses $<, >, \le, \ge$

⊕ Coordinate Plane

Ordered pair: (x, y)

x-axis: horizontal y-axis: vertical

Origin: $(0, 0)$

Four quadrants (I, II, III, IV)

▦ Statistics

Mean: sum of values $\div$ count

Median: middle value (sorted)

Range: max $-$ min

Championship Scoreboard

Track your rise through every championship round

Champion's Name: ______________________________

🏆 Round	🎖 Tier	📅 Date	⭐ Score	🙂 Rating
1	Bronze		/	
2	Bronze		/	
3	Bronze		/	
4	Silver		/	
5	Silver		/	
6	Silver		/	
7	Gold		/	
8	Gold		/	
9	Gold		/	
10	👑		/	

👑 Champion's Reflection

My strongest topics (where I consistently score well):

____

____

Topics I improved on the most from Bronze to Gold:

____

____

My score trend (Bronze avg → Gold avg → Championship):

____

____

One strategy that helped me improve the most:

____

____

My confidence level for the real test (1–10): _________ / 10

Find more at
ViewMath.com/CO-Grade6

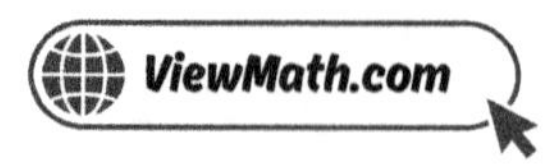

★ Table of Contents ★

Here's what we'll explore together!

 Let's learn and have fun!

Practice Test 1

 *30 Questions*

✏️ Before You Start ✏️

- ✔ **Read each question carefully** before choosing your answer.
- ✔ **Show your work** on scratch paper when you need to.
- ✔ **Skip hard questions** and come back to them later.
- ✔ **Check your answers** when you're done.
- ✔ **Take your time** — there's no rush!

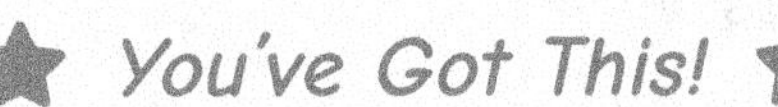

⭐ You've Got This! ⭐

Do your best and show what you know!

1. A garden has flowers and vegetables in a ratio of 4 : 1. There are 20 flowers. How many vegetables are there?

Your Answer:

2. The graph below shows the distance traveled by a delivery truck over time.

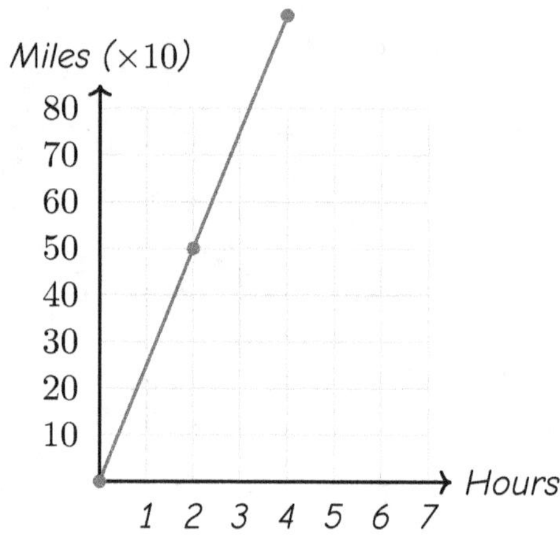

What is the truck's unit rate in miles per hour?

A) 50 miles per hour

B) 20 miles per hour

C) 25 miles per hour

D) 10 miles per hour

3. The double number line below shows equivalent ratios of scoops of mix to cups of water.

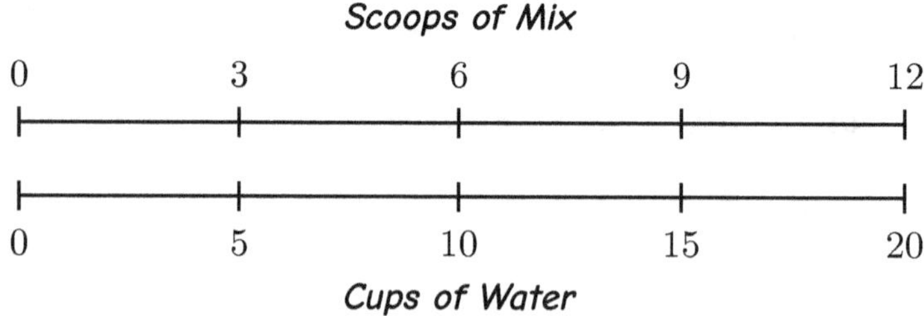

How many cups of water go with 9 scoops of mix?

A) 10

B) 12

C) 15

D) 18

Find more at
ViewMath.com/CO-Grade6

ViewMath.com

4. *A graph of a ratio relationship passes through $(0,0)$ and $(5,20)$. What is the unit rate?*

(A) 5

(B) 20

(C) 4

(D) 25

5. *The double number line below shows a map scale.*

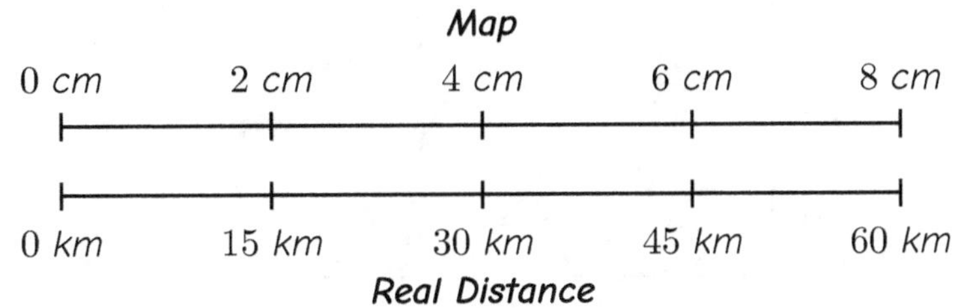

Two cities are 5 cm apart on the map. What is the real distance?

(A) 25 km

(B) 30 km

(C) 37.5 km

(D) 35 km

6. *A bottle holds 750 mL. How many full bottles can you fill from a 6-liter jug?*

Your Answer:

7. The bar graph shows the monthly earnings from different jobs for four students.

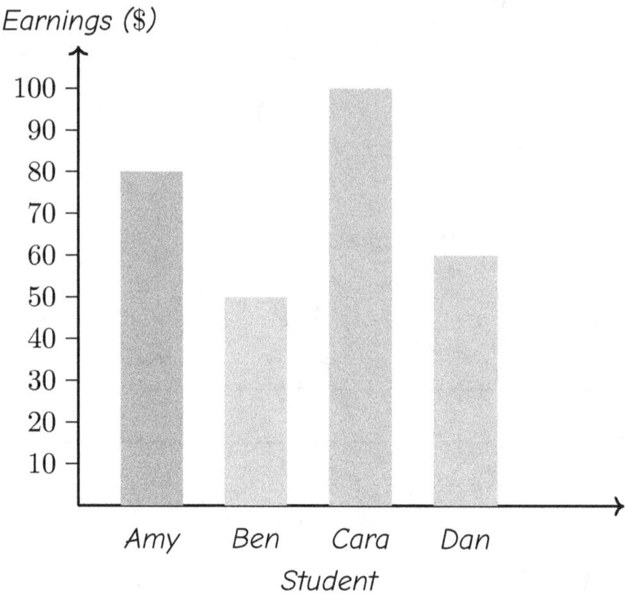

Each student saves 20% of their earnings. Which student saves the most money?

(A) Amy ($16)

(B) Ben ($10)

(C) Cara ($20)

(D) Dan ($12)

8. You deposit $500 in a savings account that earns 4% simple interest per year. How much interest do you earn after 2 years?

(A) $20

(B) $40

(C) $50

(D) $80

9. Rewrite $12 + 18 + 30$ using the GCF and a sum.

Your Answer:

10. The temperature in Anchorage, Alaska was 18 degrees below zero one morning. Write this temperature as an integer.

Your Answer

11. The number line below is divided into fourths between each pair of consecutive integers. Points A, B, C, and D are marked.

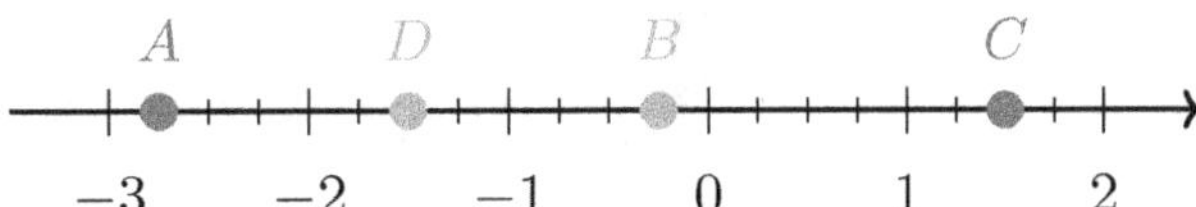

Which point represents the greatest value?

(A) A

(B) B

(C) C

(D) D

Find more at
ViewMath.com/CO-Grade6

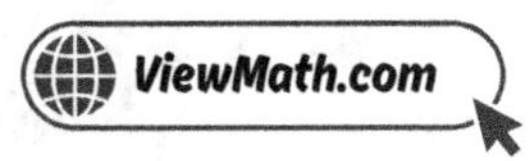

12. A town map is drawn on the coordinate grid below. Each square represents one block.

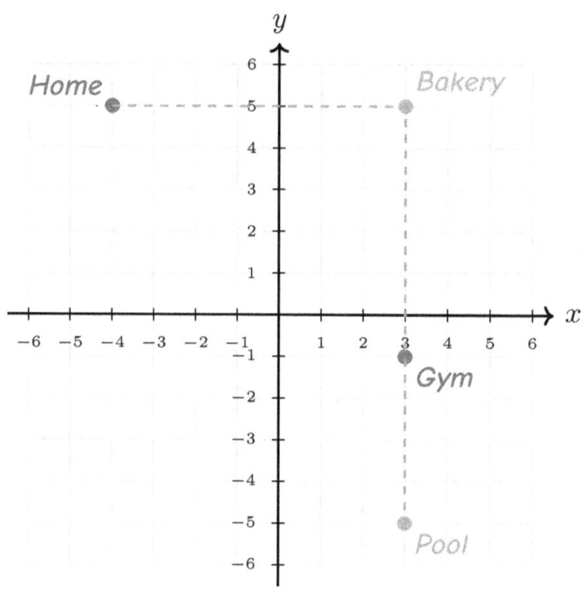

Part A: How many blocks is it from Home to the Bakery?

Part B: How many blocks is it from the Bakery to the Gym?

Part C: How many blocks is it from the Gym to the Pool?

Part D: If you walk from Home to the Bakery, then to the Gym, and finally to the Pool, what is the total distance in blocks?

Your Answer:

13. Evaluate: $5^2 + 4^2$

Your Answer:

14. Which expression represents "a number n plus 8"?

(A) $n - 8$

(B) $8n$

(C) $n + 8$

(D) $n \div 8$

15. A parking garage charges \$5 plus \$3 per hour. How much does it cost to park for 6 hours? Use the expression $5 + 3h$.

Your Answer:

16. Simplify: $2(y + 9)$

(A) $2y + 9$

(B) $2y + 11$

(C) $2y + 18$

(D) $11y$

17. Marcus has n baseball cards. He buys 12 more and then gives 5 to his friend. Write an expression for how many cards he has now.

Your Answer:

18. Solve: $\dfrac{w}{8} = 9$

Your Answer:

19. Which of the following is true about the inequality $x > -2$?

(A) $x = -2$ is a solution

(B) $x = -3$ is a solution

(C) $x = 0$ is a solution

(D) $x = -2.5$ is a solution

20. A triangular sail has a base of 3 m and a height of 7 m. How much fabric is needed to make 2 identical sails?

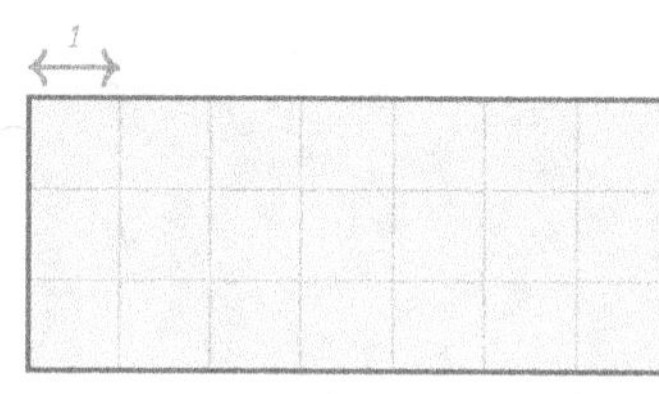

(A) $10.5\ m^2$

(B) $42\ m^2$

(C) $21\ m^2$

(D) $20\ m^2$

21. A parallelogram and a rectangle both have a base of 7 m and a height of 4 m. Which statement is true?

(A) The rectangle has a greater area.

(B) The parallelogram has a greater area.

(C) They have the same area.

(D) You cannot compare them.

22. Volume is measured in which type of units?

(A) Square units (cm^2)

(B) Linear units (cm)

(C) Cubic units (cm^3)

(D) No units are needed

23. A rectangle has vertices $(1, 2)$, $(7, 2)$, $(7, -4)$, and $(1, -4)$. What is the length of the longer side?

(A) 5 units

(B) 6 units

(C) 7 units

(D) 8 units

24. A composite figure is made of a 5×4 rectangle and a right triangle with base 5 and height 3, placed on top. What is the total area?

(A) 27.5 square units

(B) 35 square units

(C) 20 square units

(D) 32.5 square units

25. A triangular prism has 2 triangular bases and 3 rectangular faces. How many faces does it have in total?

(A) 3

(B) 4

(C) 5

(D) 6

26. Look at the flowchart below. Which path correctly identifies a statistical question?

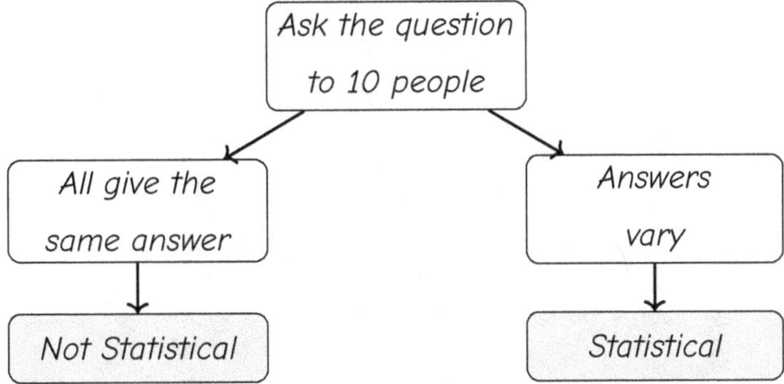

Using the flowchart, which question would end at "Statistical"?

(A) How many wheels does a bicycle have?

(B) What is the sum of $10 + 5$?

(C) How many minutes do you spend eating lunch?

(D) How many days are in one week?

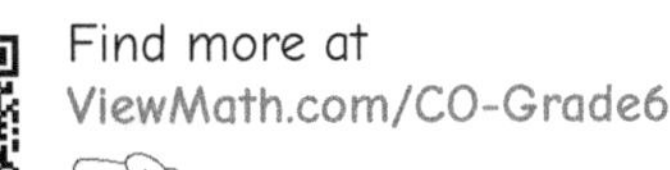
Find more at
ViewMath.com/CO-Grade6

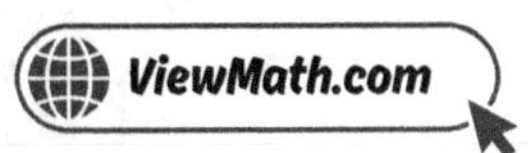

27. *Data:* $4, 5, 5, 6, 6, 6, 7, 7, 8.$ *Which feature best describes this data?*

(A) *It has a large gap.*

(B) *It has an outlier at 4.*

(C) *It clusters around 6 with a peak there.*

(D) *It is extremely spread out.*

28. *A histogram shows homework times: 0–14 min (5 students), 15–29 min (10 students), 30–44 min (8 students), 45–59 min (2 students). How many students are represented?*

Your Answer

29. *Data:* $20, 25, 30, 35, 40, 45, 50, 55, 60.$ *What is Q3?*

(A) 40

(B) 45

(C) 50

(D) 52.5

30. *A student claims: "Our class did better because our range is smaller." Is range alone enough to support this claim? Explain.*

Your Answer:

Find more at
ViewMath.com/CO-Grade6

 *End of Practice Test 1* 

Great job finishing the test!

☑ My Score

I got _____________ out of 30 questions right.

*Check your answers in the **Answer Key** at the back of the book.*

💡 *Review any questions you missed. That's how we learn!*

📊 Check Your Score Online!

Visit **ViewMath Academy** to enter your answers and see which topics you need to review. You can also explore lessons, take quizzes, track your scores, and save your progress!

viewmath.com/score/6.1.CO.16

Or go to viewmath.com/score and enter code: 6.1.CO.16

2

Practice Test 2

 30 Questions

 Before You Start

- ✓ **Read each question carefully** before choosing your answer.
- ✓ **Show your work** on scratch paper when you need to.
- ✓ **Skip hard questions** and come back to them later.
- ✓ **Check your answers** when you're done.
- ✓ **Take your time** — there's no rush!

 You've Got This! 

Do your best and show what you know!

1. Look at the tape diagram below.

Boys: ▨▨▨▨
Girls: ▢▢▢▢▢▢

There are 24 girls. How many boys are there?

(A) 4

(B) 12

(C) 16

(D) 20

2. The table shows the earnings of two babysitters.

	Hours Worked	Earnings
Lily	5	$60
Noah	4	$52

Part A: Find each babysitter's unit rate (dollars per hour).

Part B: Who earns more per hour? How much more?

Your Answer:

3. The graph below shows points that represent equivalent ratios of flour to sugar in a recipe.

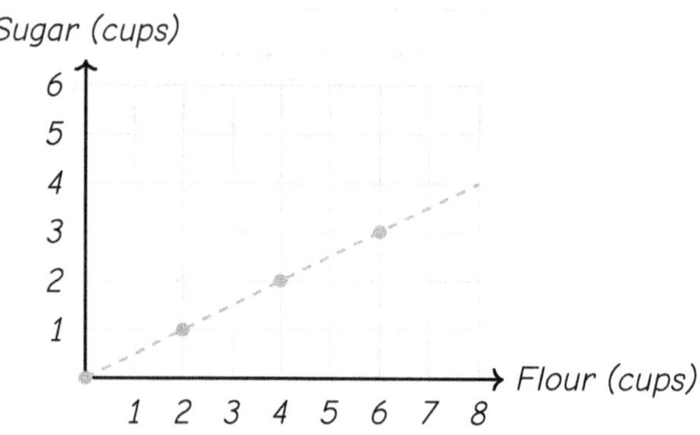

Part A: What is the ratio of flour to sugar?

Part B: If you need 8 cups of flour, how many cups of sugar do you need?

Your Answer:

4. The points $(2, 5)$ and $(4, 10)$ are on a graph. What is the ratio $x : y$?

(A) $5 : 2$

(B) $1 : 2$

(C) $2 : 5$

(D) $4 : 5$

5. A pool fills at a rate of 12 gallons per minute. How long does it take to fill 360 gallons?

(A) 25 minutes

(B) 30 minutes

(C) 36 minutes

(D) 40 minutes

6. A race is 10 kilometers. How many centimeters is that?

(A) 100,000

(B) 10,000

(C) 1,000,000

(D) 1,000

7. Mia earns $120 babysitting. She wants to save 25% of her earnings. How much should she save?

(A) $20

(B) $25

(C) $30

(D) $35

8. A family earns $3,200 per month. Their monthly expenses are: rent $960 (fixed), car payment $320 (fixed), groceries $480 (variable), entertainment $240 (variable), and utilities $320 (fixed). The rest goes to savings. What is the savings rate as a percent of income?

Your Answer

9. Which expression is equivalent to $20 + 15$ using the GCF?

(A) $5(4 + 5)$

(B) $5(4 + 3)$

(C) $3(7 + 5)$

(D) $10(2 + 5)$

10. Give a real-world example where the integer -50 could be used.

Your Answer

11. Which list shows the numbers in order from least to greatest?

$$2.5, \quad -3.1, \quad -0.8, \quad 1.2, \quad -3.5$$

(A) $-3.5, -3.1, -0.8, 1.2, 2.5$

(B) $-0.8, -3.1, -3.5, 1.2, 2.5$

(C) $-3.1, -3.5, -0.8, 1.2, 2.5$

(D) $2.5, 1.2, -0.8, -3.1, -3.5$

Find more at
ViewMath.com/CO-Grade6

12. A delivery truck drives from point $A(-4, 2)$ to point $B(3, 2)$ heading east, then turns and drives from $B(3, 2)$ to point $C(3, -5)$ heading south. What is the total distance the truck traveled?

Your Answer:

13. A student is building a cube out of unit cubes. Each edge of the cube is 3 units long. The diagram shows the base layer.

Base layer (top view)

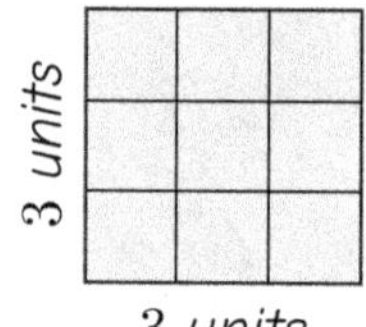

How many unit cubes are needed to build the full cube? Write your answer using an exponent.

Your Answer:

14. You earn \$12 per hour and worked h hours. You also got a \$20 bonus. Write an expression for your total pay.

Your Answer:

15. Evaluate $\dfrac{x + y}{2}$ when $x = 10$ and $y = 6$.

(A) 3

(B) 5

(C) 8

(D) 16

16. Are $4(n+2)$ and $4n+8$ equivalent?

(A) Yes — they always give the same value

(B) No — they only match when $n = 2$

(C) No — $4(n+2) = 4n+2$

(D) Yes — but only for positive values of n

17. In the formula $P = 4s$, the variable s represents the side length of a square. What does P represent?

(A) The area of the square

(B) The perimeter of the square

(C) The number of sides

(D) The diagonal of the square

18. Solve: $\dfrac{t}{6} = 5$

(A) $t = 11$

(B) $t = 1$

(C) $t = 30$

(D) $t = 56$

19. Write an inequality: A suitcase must weigh less than 50 pounds. Let w = weight.

Your Answer:

20. A triangle has base 14 m and height 8 m. What is its area?

(A) $112\ m^2$

(B) $22\ m^2$

(C) $44\ m^2$

(D) $56\ m^2$

21. A parallelogram has base 8 m and height 3.5 m. What is the area?

(A) $28\ m^2$

(B) $23\ m^2$

(C) $14\ m^2$

(D) $11.5\ m^2$

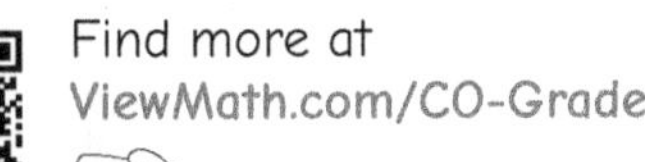

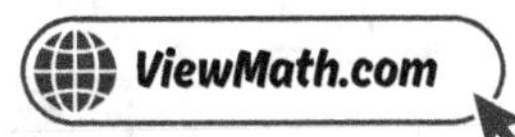

22. A swimming pool is 25 m long, 10 m wide, and 2 m deep. How many cubic meters of water does it hold when full?

Your Answer:

23. What is the distance between $(3, -1)$ and $(3, 7)$?

(A) 6 units

(B) 7 units

(C) 8 units

(D) 4 units

24. A rectangle with vertices $(1, 2)$, $(7, 2)$, $(7, 6)$, and $(1, 6)$ has a right triangle cut from it with vertices $(1, 2)$, $(7, 2)$, and $(7, 6)$. What is the area of the remaining piece?

Your Answer:

25. Which net below could fold into a cube?

A   B   C

(A) Net A only

(B) Net B only

(C) Net C only

(D) Nets A and B

26. The table below shows data collected from two different questions asked to 5 students.

Student	Question 1 Answer	Question 2 Answer
Ana	3	12
Ben	3	12
Cara	3	12
Dan	3	12
Eva	3	12

Is Question 1 statistical? Is Question 2 statistical? Explain how you can tell from the table.

Your Answer:

27. The data set $12, 14, 15, 15, 16, 17, 18$ has values that range from 12 to 18. What does this tell you about the data?

(A) The center is 15.

(B) The data is very spread out.

(C) The spread (range) is 6.

(D) The data has an outlier.

28. A dot plot shows data clustered from 20 to 25 with no dots at 26, 27, 28, 29, and one dot at 30. What feature exists between 25 and 30?

(A) A cluster

(B) A peak

(C) A gap

(D) An interval

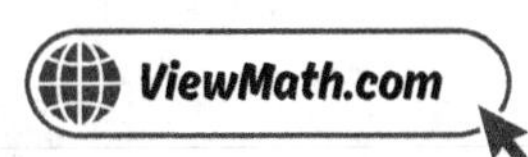

29. Look at the two box plots below showing test scores for two classes.

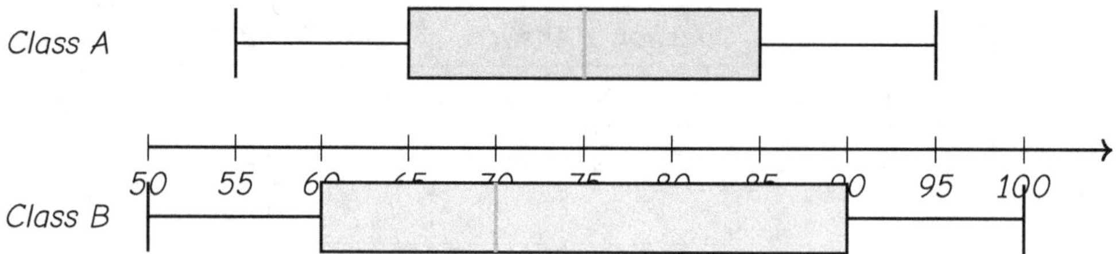

Which class has more consistent scores?

(A) Class A, because its box is narrower.

(B) Class B, because its box is wider.

(C) Both are equally consistent.

(D) Class B, because its median is higher.

30. The dot plots below show the number of sit-ups completed by students in two PE classes.

Class X

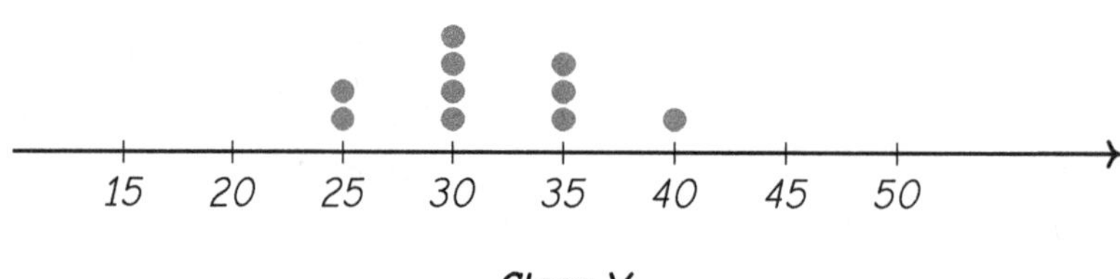

Class Y

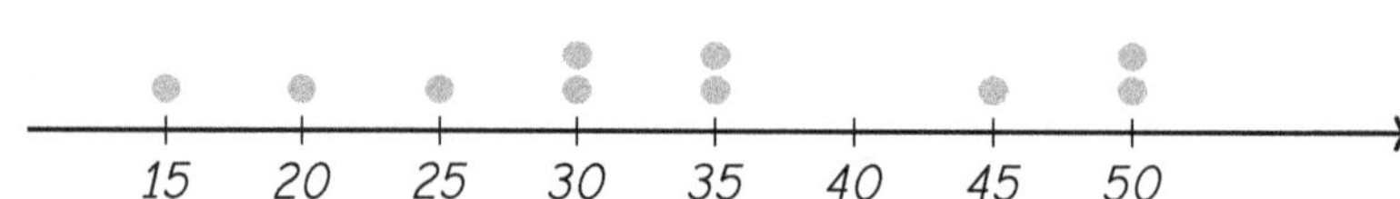

Which class has data that is more spread out?

(A) Class X

(B) Class Y

(C) They have the same spread.

(D) Cannot be determined.

Find more at
ViewMath.com/CO-Grade6

ViewMath.com

 End of Practice Test 2 

Great job finishing the test!

☑ My Score

I got ___________ out of 30 questions right.

*Check your answers in the **Answer Key** at the back of the book.*

💡 *Review any questions you missed. That's how we learn!*

📊 Check Your Score Online!

Visit **ViewMath Academy** to enter your answers and see which topics you need to review. You can also explore lessons, take quizzes, track your scores, and save your progress!

viewmath.com/score/6.1.CO.17

Or go to viewmath.com/score and enter code: 6.1.CO.17

Practice Test 3

☑ *30 Questions*

✏️ Before You Start ✏️

- ✔ **Read each question carefully** before choosing your answer.
- ✔ **Show your work** on scratch paper when you need to.
- ✔ **Skip hard questions** and come back to them later.
- ✔ **Check your answers** when you're done.
- ✔ **Take your time** — there's no rush!

⭐ **You've Got This!** ⭐

Do your best and show what you know!

1. A baker uses 2 eggs **per** 3 cups of flour. Which ratio represents flour to eggs?

  (A) $2:3$ (B) $3:5$

  (C) $3:2$ (D) $2:5$

2. A box of 12 muffins costs \$9. What is the cost per muffin?

  (A) \$1.25   (B) \$0.75

  (C) \$1.33   (D) \$3.00

3. A store sells 7 bananas for \$2. How much do 21 bananas cost?

  (A) \$4   (B) \$6

  (C) \$9   (D) \$14

4. A ratio graph passes through $(6, 2)$. What is the value of y when $x = 15$?

  (A) 3   (B) 5

  (C) 7   (D) 10

5. A store sells 3 T-shirts for \$24. How much do 10 T-shirts cost?

  (A) \$72   (B) \$80

  (C) \$70   (D) \$90

6. Convert 3 miles to feet. (1 mile = 5,280 feet)

  (A) 5,280   (B) 10,560

  (C) 15,840   (D) 21,120

7. A family earns $4,000 per month. They budget 30% for housing. How much is their housing budget?

(A) $800

(B) $1,000

(C) $1,200

(D) $1,400

8. Maria earns $600 per month. She spends $450 on expenses. How much can she save?

(A) $50

(B) $100

(C) $150

(D) $200

9. Which shows $12 + 28$ written as a product using the GCF?

(A) $2(6 + 14)$

(B) $4(3 + 7)$

(C) $4(3 + 28)$

(D) $7(2 + 4)$

10. In a city, the temperature at noon was 5°F. By midnight, the temperature was −8°F. Which statement is true?

(A) Both readings are above zero.

(B) Both readings are below zero.

(C) The noon reading is above zero, and the midnight reading is below zero.

(D) The noon reading is below zero, and the midnight reading is above zero.

11. Which statement correctly compares $-\dfrac{7}{8}$ and $-\dfrac{3}{8}$?

(A) $-\dfrac{7}{8} > -\dfrac{3}{8}$ because $7 > 3$

(B) $-\dfrac{7}{8} < -\dfrac{3}{8}$ because $-\dfrac{7}{8}$ is farther from 0

(C) $-\dfrac{7}{8} = -\dfrac{3}{8}$ because they have the same denominator

(D) $-\dfrac{7}{8} > -\dfrac{3}{8}$ because both are negative

12. **Part A:** Find the horizontal distance between $(-5, 3)$ and $(2, 3)$.

Part B: Find the vertical distance between $(2, 3)$ and $(2, -4)$.

Part C: If you walk from $(-5, 3)$ to $(2, 3)$ and then from $(2, 3)$ to $(2, -4)$, what is the total distance you walk?

Your Answer:

13. What is the value of 10^0?

(A) 0

(B) 1

(C) 10

(D) 100

14. Write an expression for: "the product of 7 and w, increased by 5."

Your Answer:

15. Evaluate $2(3y - 1)$ when $y = 5$.

(A) 28

(B) 14

(C) 29

(D) 10

16. Simplify: $5p + 3 + 2p + 4p - 1$

(A) $11p + 2$

(B) $7p + 2$

(C) $11p + 4$

(D) $12p$

17. The table below shows the cost of buying t T-shirts from an online store that charges \$4 for shipping. Which expression was used?

T-shirts (t)	Total Cost
1	\$16
2	\$28
3	\$40
4	\$52

(A) $16t$

(B) $12t + 4$

(C) $4t + 12$

(D) $14t$

18. Solve: $p - 11 = 25$

(A) $p = 14$

(B) $p = 36$

(C) $p = 275$

(D) $p = 225$

19. A library book can be checked out for at most 21 days. Which inequality models the number of days d?

(A) $d < 21$

(B) $d > 21$

(C) $d \leq 21$

(D) $d \geq 21$

20. A triangular pennant has a base of 5 in and a height of 12 in. What is its area?

(A) $60 \ in^2$

(B) $17 \ in^2$

(C) $30 \ in^2$

(D) $34 \ in^2$

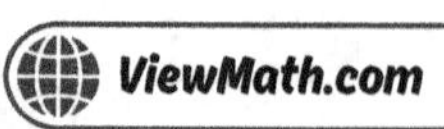

21. A trapezoid has bases 9 in and 15 in and height 6 in. What is the area?

(A) 90 in^2

(B) 72 in^2

(C) 45 in^2

(D) 135 in^2

22. An L-shaped solid is formed by joining two rectangular prisms as shown. What is the total volume?

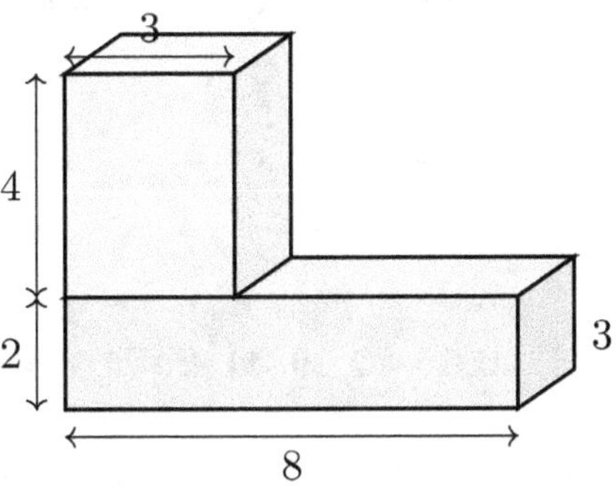

All depths are 3 units.

Your Answer:

23. A square has one vertex at $(-2, -2)$ and side length 5. The sides are horizontal and vertical. Which could be the opposite vertex?

(A) $(3, 3)$

(B) $(3, -7)$

(C) $(5, 5)$

(D) $(2, 2)$

24. A rectangle has vertices $(-5, -1)$, $(3, -1)$, $(3, 4)$, and $(-5, 4)$. Emma says the area is 30 square units, and Jake says it is 40 square units. Who is correct?

(A) Emma

(B) Jake

(C) Neither — the area is 35 square units.

(D) Neither — the area is 45 square units.

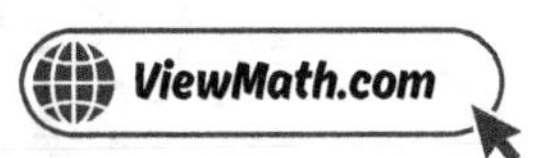

25. A rectangular prism has a surface area of 94 cm². Its length is 5 cm, width is 3 cm, and height is 4 cm. Is this correct?

    (A) Yes, the surface area is 94 cm².

    (B) No, the surface area is 120 cm².

    (C) No, the surface area is 60 cm².

    (D) No, the surface area is 47 cm².

26. Explain why "What is my teacher's age?" is **not** a statistical question.

Your Answer:

27. Data set A: 20, 21, 22, 23, 24. Data set B: 10, 15, 22, 29, 34. Both have a center near 22. What is different?

    (A) Data set A is more spread out.

    (B) Data set B is more spread out.

    (C) Both have the same spread.

    (D) Neither data set has a center.

28. A dot plot shows quiz scores: 6 (2 dots), 7 (5 dots), 8 (4 dots), 9 (3 dots), 10 (1 dot). How many students took the quiz?

    (A) 10

    (B) 15

    (C) 20

    (D) 40

29. What percent of the data falls within the box of a box plot?

    (A) 25%

    (B) 50%

    (C) 75%

    (D) 100%

Find more at
ViewMath.com/CO-Grade6

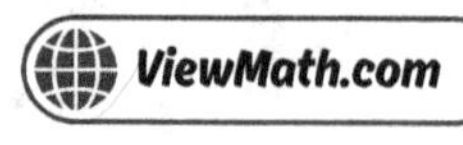

30. When two data sets have very little overlap, what can you say?

(A) The groups are very similar.

(B) The groups are clearly different from each other.

(C) Both groups have the same center.

(D) Both groups have the same spread.

 # End of Practice Test 3 

Great job finishing the test!

 **My Score**

I got _____________ out of 30 questions right.

*Check your answers in the **Answer Key** at the back of the book.*

 *Review any questions you missed. That's how we learn!*

Check Your Score Online!

Visit **ViewMath Academy** to enter your answers and see which topics you need to review. You can also explore lessons, take quizzes, track your scores, and save your progress!

viewmath.com/score/6.1.CO.18

Or go to viewmath.com/score and enter code: 6.1.CO.18

Practice Test 4

 30 Questions

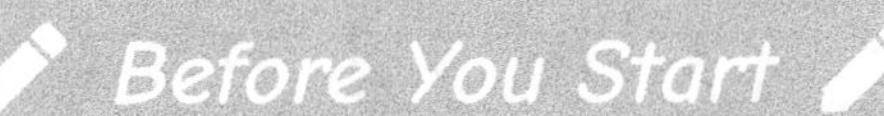

✏️ Before You Start ✏️

- ✓ **Read each question carefully** before choosing your answer.
- ✓ **Show your work** on scratch paper when you need to.
- ✓ **Skip hard questions** and come back to them later.
- ✓ **Check your answers** when you're done.
- ✓ **Take your time** — there's no rush!

★ **You've Got This!** ★

Do your best and show what you know!

1. A store sells 3 pencils **for every** 1 eraser. Which ratio represents pencils to erasers?

   (A) $1:3$ (B) $3:1$

   (C) $3:4$ (D) $1:4$

2. A bike travels at a unit rate of 14 miles per hour. How long will it take to travel 49 miles?

   Your Answer:

3. The ratio table below has an error in one row. Which row has the error?

x	y
4	10
8	20
12	25
16	40

   (A) Row 1 (B) Row 2

   (C) Row 3 (D) Row 4

4. On a graph, Line A passes through $(1, 6)$ and Line B passes through $(1, 4)$. Both go through the origin. Which line represents a faster rate? Explain.

   Your Answer:

5. A car travels 180 miles in 3 hours. At the same rate, how far will it go in 7 hours?

   (A) 360 miles (B) 420 miles

   (C) 540 miles (D) 300 miles

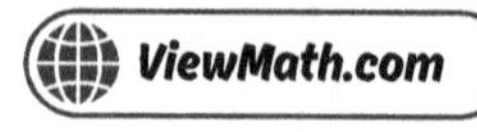

6. How many cups are in 3 quarts? (1 quart = 4 cups)

(A) 7

(B) 8

(C) 10

(D) 12

7. Marcus has \$150 on a credit card at 10% annual interest and \$150 in a savings account earning 2% annual interest. After one year with no payments or withdrawals, what is the difference between the interest he **owes** and the interest he **earns**?

(A) \$8

(B) \$12

(C) \$15

(D) \$18

8. You deposit \$600 into a savings account that earns 5% simple interest per year. How much interest do you earn after 3 years?

Your Answer:

9. Which expression shows $16 + 40$ rewritten using the GCF?

(A) $4(4 + 10)$

(B) $8(2 + 5)$

(C) $2(8 + 20)$

(D) $8(2 + 40)$

10. Which of the following statements about zero is true?

(A) Zero is a positive number.

(B) Zero is a negative number.

(C) Zero is neither positive nor negative.

(D) Zero is both positive and negative.

11. Which number is the least?

$$-1.5, \quad \frac{3}{2}, \quad -\frac{7}{4}, \quad 0$$

(A) -1.5

(B) $\frac{3}{2}$

(C) $-\frac{7}{4}$

(D) 0

12. The points $(5, -1)$ and $(-3, -1)$ lie on a horizontal line. Explain step by step how to find the distance between them, and state why the answer is positive.

Your Answer:

13. The table below shows powers of 3. What value belongs in the blank?

3^1	3^2	3^3	3^4	3^5
3	9	27	?	243

(A) 36

(B) 64

(C) 81

(D) 108

14. Write an expression for: "a number t divided by 4, then subtract 1."

Your Answer:

15. *Evaluate* $k^2 + k$ *when* $k = 3$.

(A) 6

(B) 9

(C) 12

(D) 15

16. *The diagram shows two groups. Which expression represents the total?*

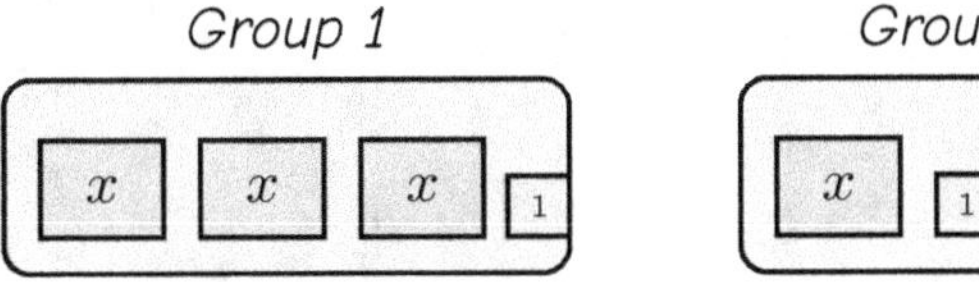

(A) $3x + 4$

(B) $4x + 4$

(C) $4x + 3$

(D) $3x + 1 + x + 3$

17. *In the formula* $d = 60t$, *what could d and t represent?*

(A) d = days, t = hours

(B) d = distance in miles, t = time in hours

(C) d = dollars, t = tax rate

(D) d = degrees, t = temperature

18. *Solve:* $x + 3.5 = 10$

(A) $x = 6.5$

(B) $x = 13.5$

(C) $x = 7.5$

(D) $x = 3.5$

19. *Which of the following values is NOT a solution to* $y \geq 3$?

(A) $y = 3$

(B) $y = 4$

(C) $y = 100$

(D) $y = 2.5$

20. *A triangle has base 20 in and height 7 in. What is the area?*

(A) 140 in^2

(B) 27 in^2

(C) 70 in^2

(D) 54 in^2

21. *A parallelogram has base 12 m and a slant side of 8 m. The height is 6 m. What is the area?*

(A) 96 m^2

(B) 72 m^2

(C) 48 m^2

(D) 36 m^2

22. *A rectangular prism has length $\frac{1}{2}$ m, width $\frac{1}{2}$ m, and height $\frac{1}{2}$ m. What is the volume?*

(A) $\frac{1}{8}$ m^3

(B) $\frac{1}{4}$ m^3

(C) $\frac{3}{2}$ m^3

(D) $\frac{1}{2}$ m^3

23. *Look at the triangle plotted below. What is the length of side $\overline{AB}$?*

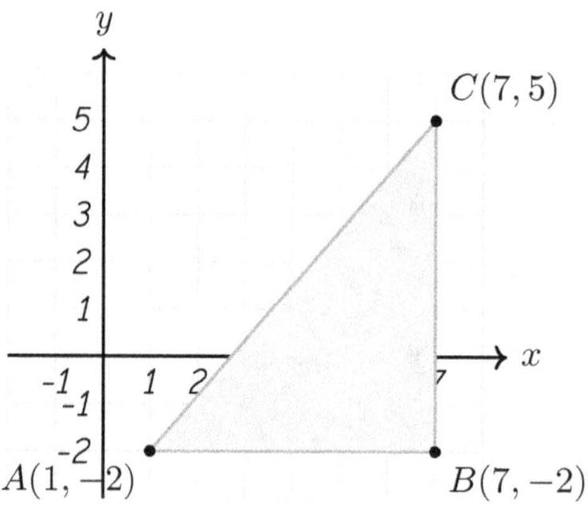

(A) 5 units

(B) 6 units

(C) 7 units

(D) 8 units

24. An L-shaped figure has vertices $(0,0)$, $(10,0)$, $(10,4)$, $(6,4)$, $(6,8)$, and $(0,8)$. What is the total area?

Your Answer:

25. What is a net?

(A) A 3D shape made of cubes

(B) A flat pattern that folds into a 3D shape

(C) The volume of a rectangular prism

(D) A grid used to measure area

26. A scientist asks, "What is the wingspan of each bald eagle in the sanctuary?" This is statistical because —

(A) it is about science

(B) the wingspan of each eagle may be different

(C) bald eagles are interesting

(D) wingspans are measured in centimeters

27. Student test times (minutes): $8, 9, 10, 10, 11, 12, 25$. A classmate says the typical time is about 12 minutes. Is this a good estimate?

(A) Yes, 12 is in the data set.

(B) Yes, because the outlier should be included.

(C) No, most times cluster around 9–11, so a typical value is closer to 10.

(D) No, the typical value is 25.

28. A histogram has the following bars: 0–9 (height 3), 10–19 (height 7), 20–29 (height 10), 30–39 (height 5). How many data values are there in total?

(A) 4

(B) 10

(C) 25

(D) 39

29. The two box plots below represent daily sales (in dollars) at two stores over the same month.

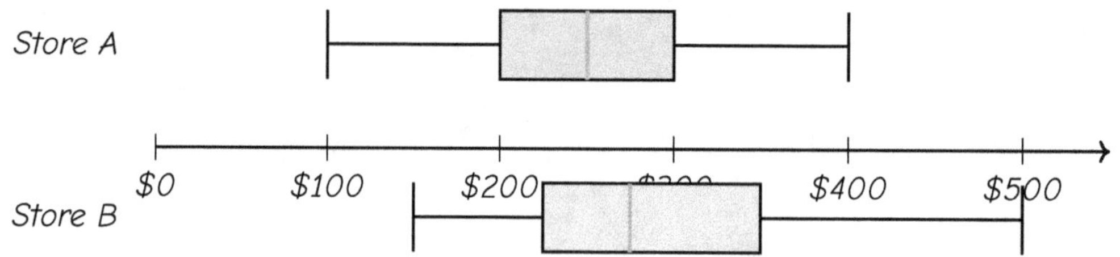

Compare the two stores. Which store has higher typical sales? Which has more predictable sales? Use the box plots to support your answers.

Your Answer:

30. Runner A's race times (min): mean $= 25$, MAD $= 1$. Runner B's times: mean $= 24$, MAD $= 5$. Which runner is faster on average? Which is more consistent?

(A) Runner A is faster and more consistent.

(B) Runner B is faster but Runner A is more consistent.

(C) Runner B is faster and more consistent.

(D) They are the same speed.

 End of Practice Test 4 

Great job finishing the test!

 My Score

I got _____________ out of 30 questions right.

*Check your answers in the **Answer Key** at the back of the book.*

💡 *Review any questions you missed. That's how we learn!*

📊 Check Your Score Online!

Visit **ViewMath Academy** to enter your answers and see which topics you need to review. You can also explore lessons, take quizzes, track your scores, and save your progress!

viewmath.com/score/6.1.CO.19

Or go to viewmath.com/score and enter code: 6.1.CO.19

Practice Test 5

 *30 Questions*

 Before You Start 

- ✓ **Read each question carefully** before choosing your answer.
- ✓ **Show your work** on scratch paper when you need to.
- ✓ **Skip hard questions** and come back to them later.
- ✓ **Check your answers** when you're done.
- ✓ **Take your time** — there's no rush!

 You've Got This! 

Do your best and show what you know!

1. A class votes on two activities: hiking and swimming. The ratio of votes for hiking to swimming is 3 : 2. There are 25 votes total. How many voted for swimming?

Your Answer:

2. A car travels 240 miles using 8 gallons of gas. What is the unit rate in miles per gallon?

(A) 8

(B) 32

(C) 30

(D) 248

3. It takes 6 cups of flour to make 4 loaves of bread. How many cups of flour for 10 loaves?

(A) 12

(B) 10

(C) 15

(D) 20

Find more at
ViewMath.com/CO-Grade6

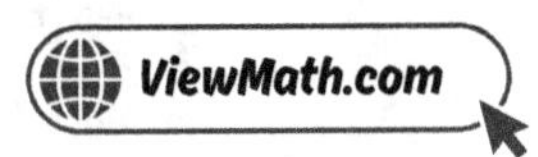

4. *The table and partial graph below show a ratio relationship.*

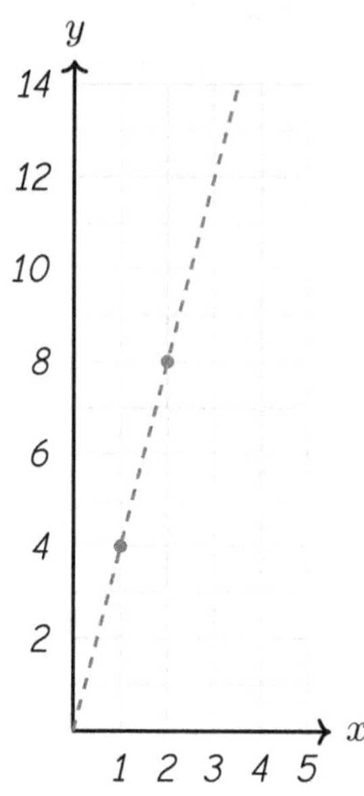

x	y
1	4
2	8
3	?

What is the missing value of y when $x = 3$?

(A) 10

(B) 12

(C) 14

(D) 16

5. *A map scale says 1 cm = 5 km. Two cities are 8 cm apart on the map. What is the real distance?*

(A) 13 km

(B) 5 km

(C) 40 km

(D) 80 km

6. *Convert 2 hours, 30 minutes to minutes, then to seconds.*

Your Answer:

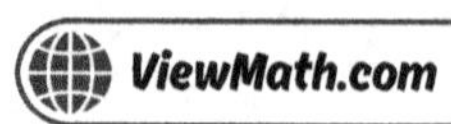

7. When you use a **debit card**, the money comes directly from your —

(A) credit card company

(B) the store's account

(C) a loan from the bank

(D) your own bank account

8. Mia earns $1,000 per month. Her budget: rent 35%, utilities 10%, food 20%, transportation 15%, and the rest goes to savings. She gets a raise to $1,200 per month but keeps the same percentages. How much **more** does she save per month after the raise?

(A) $20

(B) $30

(C) $40

(D) $50

9. Use the distributive property to find 5×46.

(A) 230

(B) 220

(C) 250

(D) 210

10. Carlos earned $20 doing chores and then spent $20 at the store. Write an integer to describe his overall change in money.

Your Answer:

11. Order the following numbers from least to greatest:

$$-\frac{2}{3}, \quad 0.5, \quad -1.25, \quad \frac{3}{4}, \quad -0.1$$

Your Answer:

Find more at
ViewMath.com/CO-Grade6

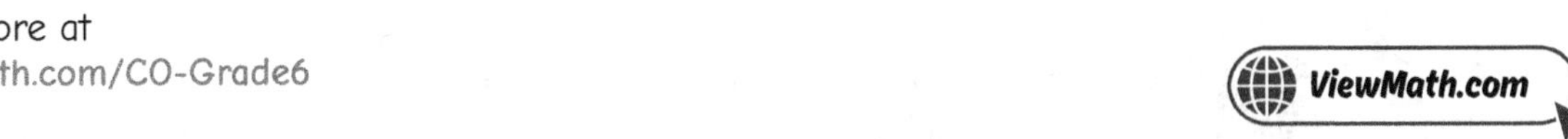

12. What is the distance between $(4, -6)$ and $(4, -1)$?

    (A) -7 (B) -5

    (C) 5 (D) 7

13. Evaluate: $(12 - 4)^2 \div 16$

    Your Answer

14. Sam has x stickers. He gives away 6. Which expression shows how many stickers Sam has now?

    (A) $x + 6$ (B) $6x$

    (C) $6 - x$ (D) $x - 6$

15. Evaluate $20 - 3n$ when $n = 4$.

    (A) 8 (B) 12

    (C) 17 (D) 32

16. Are $5(x + 2)$ and $5x + 10$ equivalent? Show a test with $x = 3$.

    Your Answer

17. In $P = 4s$, can s be any positive number, or is it one specific number?

    (A) Only $s = 1$ (B) Only whole numbers

    (C) Any positive number — the formula works for all squares   (D) Only $s = 4$

Find more at
ViewMath.com/CO-Grade6

18. Solve: $y - 8 = 14$

(A) $y = 6$

(B) $y = 22$

(C) $y = 18$

(D) $y = 112$

19. Which symbol correctly completes the statement? "The speed limit is 65 mph. You must drive _______ 65 mph."

(A) $>$

(B) $<$

(C) $\leq$

(D) $\geq$

20. Two triangles both have a base of 10 cm. Triangle P has a height of 6 cm and Triangle Q has a height of 8 cm. How much greater is the area of Triangle Q?

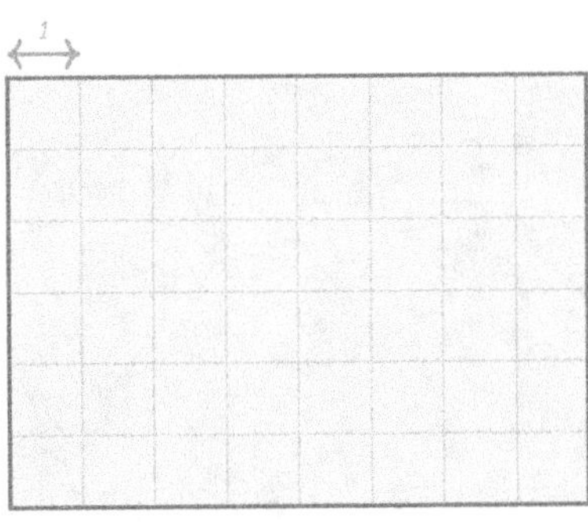

(A) $2\ cm^2$

(B) $10\ cm^2$

(C) $20\ cm^2$

(D) $40\ cm^2$

21. A trapezoid has an area of $84\ cm^2$ and bases of 10 cm and 14 cm. What is the height?

Your Answer:

Find more at
ViewMath.com/CO-Grade6

22. A rectangular prism has a volume of 120 cm^3. Its length is 10 cm and width is 4 cm. What is the height?

(A) 3 cm

(B) 12 cm

(C) 6 cm

(D) 30 cm

23. A rectangle on the coordinate plane has a length of 10 units and width of 4 units. One vertex is at the origin $(0,0)$ and the sides are along the axes. Which of the following could be the opposite vertex?

(A) $(10, 4)$

(B) $(14, 0)$

(C) $(4, 4)$

(D) $(10, 10)$

24. A rectangle has vertices $(0,0)$, $(6,0)$, $(6,4)$, and $(0,4)$. What is the area?

(A) 10 square units

(B) 20 square units

(C) 24 square units

(D) 12 square units

25. Surface area is measured in:

(A) Cubic units

(B) Linear units

(C) Square units

(D) No units

26. Which question would produce data where every answer is the same?

(A) How many hours do you watch TV each day?

(B) How many feet are in one yard?

(C) What is your favorite sport?

(D) How far can you throw a ball?

27. Data: $3, 4, 5, 5, 6, 6, 6, 7, 7, 8$. Where does the data cluster? What is the peak?

Your Answer:

Find more at
ViewMath.com/CO-Grade6

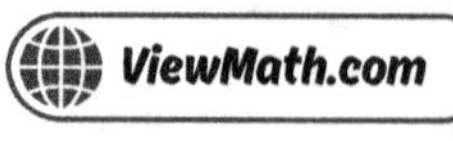

28. A dot plot has most of its dots between 40 and 50, with one dot at 10 and one at 80. What describes this data?

(A) Symmetric with no outliers

(B) Clustered near 40–50 with possible outliers at 10 and 80

(C) Uniform with equal spread

(D) Has a gap but no cluster

29. The data table below shows the number of books 9 students read this semester.

Student	1	2	3	4	5	6	7	8	9
Books	2	4	5	7	8	9	10	12	15

Find the five-number summary and the IQR.

Your Answer:

30. Data: $10, 12, 14, 16, 18, 20, 22$. The mean is 16. A student says the data is "very spread out." Is this correct?

(A) Yes, the range is 12.

(B) No, the range (12) is moderate and the values increase by a steady 2.

(C) Yes, because there are 7 values.

(D) No, because the mean equals the median.

 End of Practice Test 5 

Great job finishing the test!

 My Score

I got _____________ out of 30 questions right.

Check your answers in the **Answer Key** at the back of the book.

Review any questions you missed. That's how we learn!

Check Your Score Online!

Visit **ViewMath Academy** to enter your answers and see which topics you need to review. You can also explore lessons, take quizzes, track your scores, and save your progress!

viewmath.com/score/6.1.CO.20

Or go to viewmath.com/score and enter code: 6.1.CO.20

Practice Test 6

 30 Questions

✏ Before You Start ✏

- ✔ **Read each question carefully** before choosing your answer.
- ✔ **Show your work** on scratch paper when you need to.
- ✔ **Skip hard questions** and come back to them later.
- ✔ **Check your answers** when you're done.
- ✔ **Take your time** — there's no rush!

★ You've Got This! ★

Do your best and show what you know!

1. A smoothie recipe uses 3 bananas **for every** 4 cups of yogurt. Write the ratio of bananas to yogurt. Then write the ratio of yogurt to bananas.

Your Answer

2. Brand X sells 5 notebooks for $8.75. Brand Y sells 3 notebooks for $4.50. Which is the better deal?

(A) Brand X at $1.75 each

(B) Brand Y at $1.50 each

(C) Brand X at $1.50 each

(D) Brand Y at $1.75 each

3. A teacher hands out 3 pencils for every 2 students. There are 16 students. How many pencils does she need?

(A) 18

(B) 24

(C) 20

(D) 32

4. A car travels at 30 miles per hour. Which set of points represents this ratio?

(A) $(1, 30), (2, 60), (3, 90)$

(B) $(30, 1), (60, 2), (90, 3)$

(C) $(1, 30), (2, 50), (3, 90)$

(D) $(1, 3), (2, 6), (3, 9)$

5. Which strategy is best for solving "how much for one?" problems?

(A) Tape diagram

(B) Unit rate

(C) Ratio table

(D) Equation

6. Convert 7,000 grams to kilograms.

(A) 0.7 kg

(B) 7 kg

(C) 70 kg

(D) 700 kg

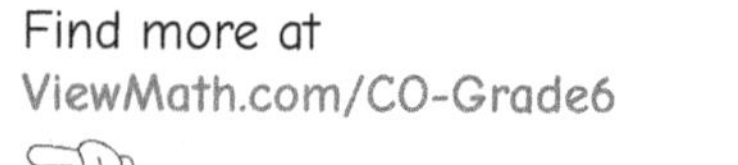

Find more at
ViewMath.com/CO-Grade6

ViewMath.com

7. Leo puts $600 in a savings account that earns 5% simple interest per year. How much total money will be in the account after 3 years?

(A) $630

(B) $690

(C) $750

(D) $900

8. A student earns $120 per week. She puts 25% into savings. After 6 weeks she deposits all her savings into an account that earns 3% simple interest per year. How much interest does she earn in 1 year?

(A) $3.60

(B) $5.40

(C) $9.00

(D) $18.00

9. Which expression shows $24 + 18$ rewritten using the GCF and the distributive property?

(A) $2(12 + 9)$

(B) $3(8 + 6)$

(C) $6(4 + 3)$

(D) $6(4 + 6)$

10. Maria has $0 in her bank account. What does this mean?

(A) She owes money.

(B) She has some savings.

(C) She has no money and no debt.

(D) She has negative money.

11. Which pair of numbers is in order from least to greatest?

(A) $-\dfrac{1}{2}, \ -\dfrac{3}{4}$

(B) $-0.9, \ -0.6$

(C) $-2.1, \ -2.8$

(D) $\dfrac{1}{4}, \ -\dfrac{1}{4}$

12. *Two points are plotted on the coordinate plane below. What is the distance between them?*

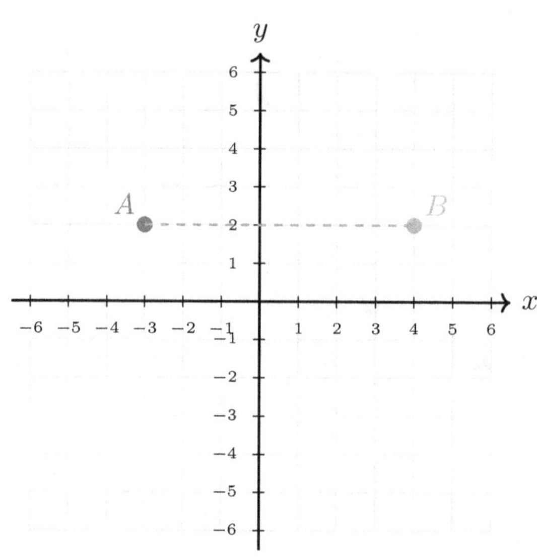

(A) 2 units

(B) 4 units

(C) 5 units

(D) 7 units

---

13. *Evaluate:* $60 \div (3 + 2) - 2^2$

Your Answer:

---

14. *Which expression represents "divide 20 by a number p"?*

(A) $20 + p$

(B) $p \div 20$

(C) $20p$

(D) $20 \div p$

---

15. *Evaluate* $\dfrac{n}{3} + 8$ *when* $n = 15.$

(A) 5

(B) 11

(C) 13

(D) 23

16. Which expression is equivalent to $10a - 2a + 7$?

(A) $12a + 7$

(B) $8a + 7$

(C) $8a - 7$

(D) $15a$

17. A dog walker earns \$7 per dog. Write an expression for the earnings from walking d dogs, and find the earnings from walking 8 dogs.

Your Answer:

18. Five friends share the cost of a gift equally. Each friend pays \$9. What is the total cost of the gift?

Your Answer:

19. Which inequality represents "a number y is fewer than 20"?

(A) $y > 20$

(B) $y \geq 20$

(C) $y < 20$

(D) $y \leq 20$

20. A triangle has base $\frac{3}{4}$ ft and height $\frac{2}{3}$ ft. What is its area?

Your Answer:

21. A trapezoid has bases $b_1 = 4$ cm and $b_2 = 4$ cm with height 6 cm. What shape is this trapezoid actually equivalent to?

(A) A triangle

(B) A parallelogram

(C) A circle

(D) A pentagon

Find more at
ViewMath.com/CO-Grade6

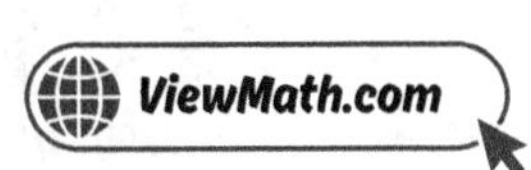

22. A rectangular prism has volume 180 cm^3, length 9 cm, and width 5 cm. What is the height?

23. Which pair of points forms a vertical segment?

(A) $(3, 2)$ and $(7, 2)$

(B) $(5, -1)$ and $(5, 4)$

(C) $(1, 3)$ and $(4, 6)$

(D) $(0, 0)$ and $(3, 0)$

24. To find the area of an irregular polygon on the coordinate plane, you can:

(A) Multiply all the coordinates together.

(B) Break it into rectangles and triangles, then add the areas.

(C) Count only the vertices.

(D) Subtract the perimeter from the largest coordinate.

25. Mia paints all sides of a wooden block that is 3 ft by 2 ft by 4 ft. She then paints an identical block. How much total area does she paint?

(A) 52 ft^2

(B) 104 ft^2

(C) 48 ft^2

(D) 26 ft^2

26. "What is the average rainfall in our city each month?" Is this a statistical question?

(A) No, because it asks for an average.

(B) No, because rainfall is always the same.

(C) Yes, because rainfall varies from month to month.

(D) Yes, because it mentions a city.

Find more at
ViewMath.com/CO-Grade6

27. Which describes the data set $50, 50, 50, 50, 50$?

(A) Symmetric with a large spread

(B) Symmetric with zero spread

(C) Skewed right

(D) Has an outlier

28. A dot plot of daily temperatures shows: 70°F (1 dot), 72°F (3 dots), 74°F (5 dots), 76°F (4 dots), 78°F (2 dots). What is the range of the data?

(A) 4

(B) 6

(C) 8

(D) 15

29. A box plot has $min = 10$, $Q1 = 20$, $median = 30$, $Q3 = 40$, $max = 50$. What is the IQR?

(A) 10

(B) 20

(C) 30

(D) 40

30. A good data summary should include —

(A) only the mean

(B) only the range

(C) both a measure of center and a measure of spread

(D) only the number of data values

Find more at
ViewMath.com/CO-Grade6

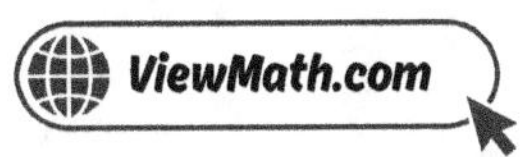

 # End of Practice Test 6 

Great job finishing the test!

 My Score

I got _____________ out of 30 questions right.

*Check your answers in the **Answer Key** at the back of the book.*

💡 *Review any questions you missed. That's how we learn!*

📊 Check Your Score Online!

Visit **ViewMath Academy** to enter your answers and see which topics you need to review. You can also explore lessons, take quizzes, track your scores, and save your progress!

viewmath.com/score/6.1.CO.21

Or go to *viewmath.com/score* and enter code: 6.1.CO.21

Practice Test 7

📋 30 Questions

✏️ **Before You Start** ✏️

- ✓ **Read each question carefully** before choosing your answer.
- ✓ **Show your work** on scratch paper when you need to.
- ✓ **Skip hard questions** and come back to them later.
- ✓ **Check your answers** when you're done.
- ✓ **Take your time** — there's no rush!

⭐ **You've Got This!** ⭐

Do your best and show what you know!

1. A lemonade stand sells 7 cups of lemonade **for each** 2 cups of iced tea. If they sold 21 cups of lemonade, how many cups of iced tea did they sell?

Your Answer:

2. A garden hose fills a 150-gallon tank in 5 hours. What is the unit rate?

(A) 25 gallons per hour

(B) 30 gallons per hour

(C) 35 gallons per hour

(D) 750 gallons per hour

3. Complete the ratio table for the ratio $5 : 8$.

5	8
10	?
?	32

Your Answer:

4. A line passes through the origin and the point $(5, 15)$. What is the ratio x to y?

(A) $1 : 5$

(B) $5 : 1$

(C) $1 : 3$

(D) $3 : 1$

5. Two families split a \$350 dinner bill in a $3 : 4$ ratio. How much does each family pay?

Your Answer:

Find more at
ViewMath.com/CO-Grade6

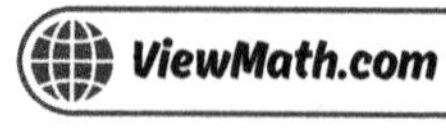

6. How many inches are in 5 feet?

(A) 50

(B) 55

(C) 60

(D) 72

7. Lily receives \$60 for her birthday. She puts 40% in savings, spends 25% on a book, and donates 10% to charity. How much money does Lily have left?

Your Answer:

8. Lily saves \$50 per month. Her monthly income is \$250. What is her savings rate?

(A) 10%

(B) 15%

(C) 20%

(D) 25%

9. Which expression is equivalent to $3(7 + 4)$?

(A) $3 \times 7 + 4$

(B) $3 \times 7 + 3 \times 4$

(C) $3 + 7 \times 4$

(D) $3 \times 7 \times 4$

10. A helicopter is flying at 1,200 feet above sea level. Which integer best describes its altitude?

(A) $-1{,}200$

(B) 0

(C) $1{,}200$

(D) -12

11. Place the following numbers in order from greatest to least:

$$-\frac{1}{2}, \quad -0.4, \quad -\frac{3}{5}, \quad 0, \quad -0.55$$

Your Answer:

12. What is the distance between $(-4, 1)$ and $(2, 1)$?

(A) -6

(B) -2

(C) 2

(D) 6

13. Evaluate: 2×3^2

(A) 12

(B) 18

(C) 36

(D) 64

14. Which expression represents "a number r squared, minus 10"?

(A) $2r - 10$

(B) $r^2 - 10$

(C) $(r - 10)^2$

(D) $r - 10^2$

15. Evaluate $8y - 3$ when $y = 6$.

Your Answer:

16. *Simplify:* $7n + 4n$

(A) $11n$

(B) $28n$

(C) $11n^2$

(D) $74n$

17. *A gym charges $25 to join and $10 per month. Which expression gives the total cost after m months?*

(A) $25m + 10$

(B) $25 + 10m$

(C) $35m$

(D) $25 - 10m$

18. *Which inverse operation would you use to solve* $\dfrac{n}{3} = 12$?

(A) Divide both sides by 3

(B) Subtract 3 from both sides

(C) Add 3 to both sides

(D) Multiply both sides by 3

19. *A movie is rated PG-13, meaning you must be over 13 to watch without a parent. Which inequality represents ages that can watch without a parent?*

(A) $a < 13$

(B) $a > 13$

(C) $a \leq 13$

(D) $a = 13$

20. *A triangular flower bed has a base of 4.5 m and a height of 6 m. What is its area?*

Your Answer:

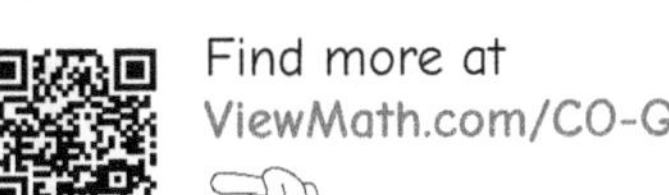

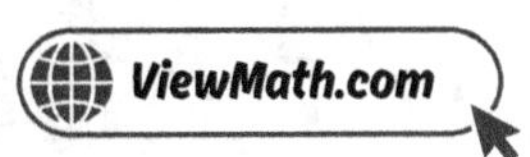

21. A park is shaped like a trapezoid with parallel sides of 40 m and 60 m and a height of 30 m. What is the area of the park?

(A) 1,800 m^2

(B) 1,200 m^2

(C) 2,400 m^2

(D) 1,500 m^2

22. A fish tank is 20 in long, 10 in wide, and 12 in tall. What is the volume of the tank?

(A) 42 in^3

(B) 240 in^3

(C) 1,200 in^3

(D) 2,400 in^3

23. What is the distance between $(-7, 3)$ and $(5, 3)$?

Your Answer:

24. What is the area of the shaded right triangle on the coordinate plane?

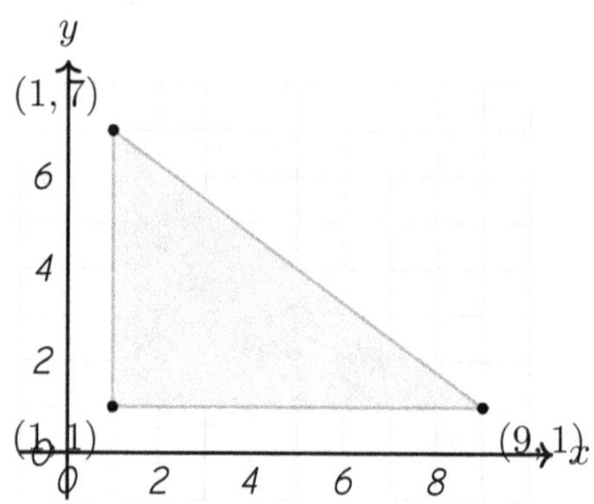

(A) 48 square units

(B) 24 square units

(C) 36 square units

(D) 12 square units

25. A classroom has dimensions 10 m by 8 m by 3 m. What is the total surface area of the walls, floor, and ceiling?

Your Answer:

26. A librarian asks, "How many books were checked out today?" Is this a statistical question?

(A) Yes, because it is about books.

(B) Yes, if she plans to ask the same question over many days.

(C) No, because on one specific day there is only one answer.

(D) No, because libraries always have the same number of checkouts.

27. The histogram below shows the number of minutes students spent reading yesterday.

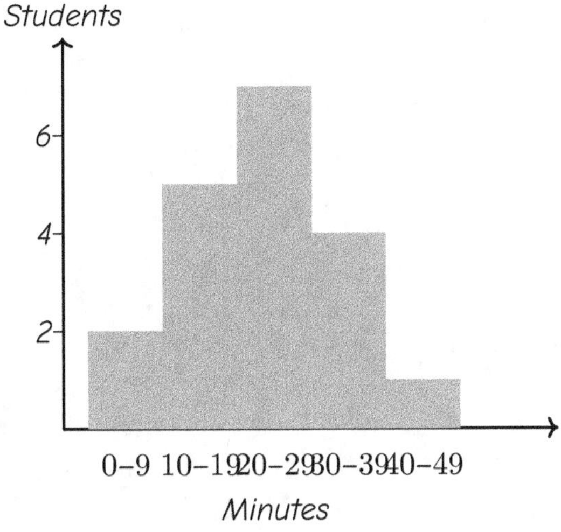

How many students were surveyed in total? Describe one feature of the data's shape.

Your Answer:

28. Which type of data display shows every individual data value?

(A) Histogram

(B) Circle graph

(C) Dot plot

(D) Box plot

29. If the whiskers of a box plot are both short and the box is narrow, what does this tell you?

(A) The data is very spread out.

(B) The data is highly consistent with little variability.

(C) There are many outliers.

(D) The data set is very large.

30. Group P: median $= 45$, IQR $= 20$. Group Q: median $= 60$, IQR $= 5$. Which group performed higher and which is more consistent?

(A) Group P is higher and more consistent.

(B) Group Q is higher and more consistent.

(C) Group P is higher, Group Q is more consistent.

(D) Group Q is higher, Group P is more consistent.

 # End of Practice Test 7 

Great job finishing the test!

✅ My Score

I got ____________ out of 30 questions right.

Check your answers in the **Answer Key** at the back of the book.

💡 Review any questions you missed. That's how we learn!

📊 Check Your Score Online!

Visit **ViewMath Academy** to enter your answers and see which topics you need to review. You can also explore lessons, take quizzes, track your scores, and save your progress!

viewmath.com/score/6.1.CO.22

Or go to **viewmath.com/score** and enter code: 6.1.CO.22

Practice Test 8

☑ 30 Questions

✏ Before You Start ✏

- ✔ **Read each question carefully** before choosing your answer.
- ✔ **Show your work** on scratch paper when you need to.
- ✔ **Skip hard questions** and come back to them later.
- ✔ **Check your answers** when you're done.
- ✔ **Take your time** — there's no rush!

⭐ You've Got This! ⭐

Do your best and show what you know!

1. The ratio of cats to dogs at a pet store is $3 : 4$. Each part in the tape diagram represents 2 animals. How many dogs are there?

(A) 6

(B) 8

(C) 3

(D) 4

2. The graph shows the number of laps a swimmer completes over time.

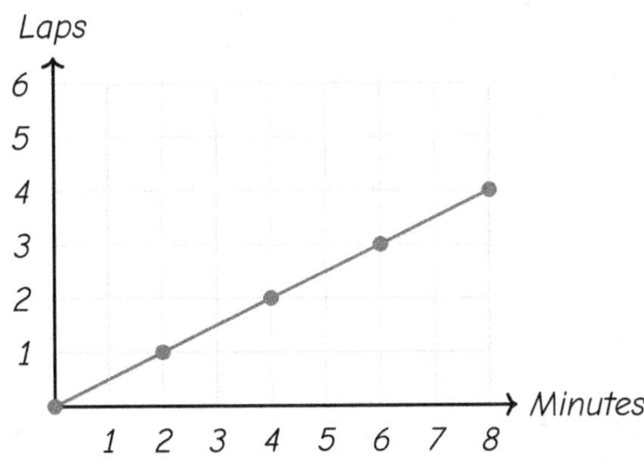

Part A: What is the swimmer's unit rate in laps per minute?
Part B: How many laps will the swimmer complete in 14 minutes?

Your Answer:

3. A smoothie recipe uses 2 cups of strawberries for every 3 cups of yogurt. How many cups of yogurt are needed for 10 cups of strawberries?

Your Answer:

4. A ratio graph passes through $(0,0)$ and $(4,7)$. What is y when $x = 8$?

(A) 11

(B) 14

(C) 15

(D) 12

Find more at
ViewMath.com/CO-Grade6

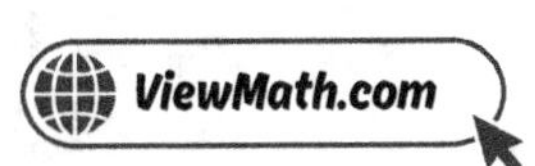

5. A printing company prints 250 flyers in 10 minutes. How long will it take to print 1,000 flyers?

Your Answer

6. The bar model below compares different length units.

1 yard		
1 ft	1 ft	1 ft

If a rope is 7 yards long, how many feet is it?

(A) 10

(B) 14

(C) 21

(D) 28

7. David deposits $1,000 in a savings account earning 3% simple interest per year. How much interest does he earn after 4 years?

(A) $30

(B) $40

(C) $120

(D) $130

8. Two friends each start saving. Plan A saves $60 per month. Plan B saves $75 per month. After 4 months, how much **more** has Plan B saved than Plan A?

(A) $15

(B) $30

(C) $45

(D) $60

9. Which expression uses the GCF to rewrite $30 + 45$?

(A) $5(6 + 9)$

(B) $15(2 + 3)$

(C) $3(10 + 15)$

(D) $15(2 + 45)$

Find more at
ViewMath.com/CO-Grade6

ViewMath.com

10. An elevator starts at ground level (0). It goes down 4 floors. Write an integer to describe where the elevator is now.

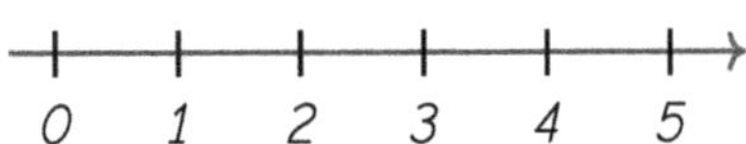

Your Answer:

11. The two thermometers below show the temperatures in two cities.

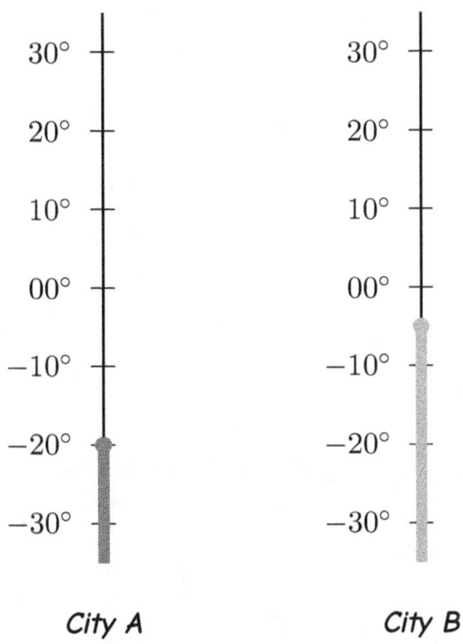

City A   City B

Which statement is true?

(A) City A is warmer because its thermometer shows a larger number

(B) City B is warmer because $-5° > -20°$

(C) City A is warmer because $20 > 5$

(D) Both cities have the same temperature

12. A rectangle has two vertices at $(1, 3)$ and $(1, -2)$. What is the length of the side that connects them?

(A) 1

(B) 3

(C) 4

(D) 5

Find more at
ViewMath.com/CO-Grade6

ViewMath.com

13. Insert one pair of parentheses to make this equation true: $2 + 3 \times 4 = 20$.

Your Answer:

14. Look at the two-step diagram below. Write the expression that describes the final output when starting with the input n.

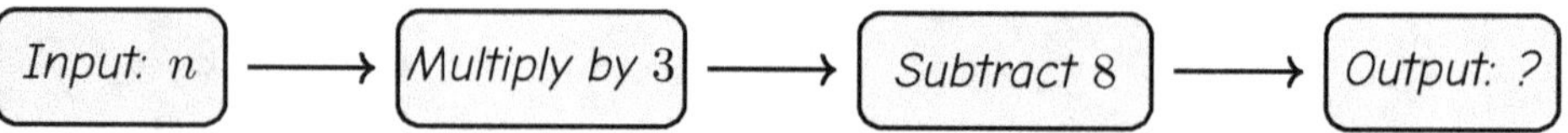

Your Answer:

15. Evaluate $m^2 + 2m + 1$ when $m = 4$.

Your Answer:

16. A student simplified $4(2x + 3) = 8x + 3$. Find and fix the error.

Your Answer:

17. A phone battery starts at 100% and loses 8% each hour. Write an expression for the battery level after h hours.

Your Answer:

18. Is $x = 5$ a solution to $3x = 18$?

(A) Yes, because $3 + 5 = 18$

(B) Yes, because $3 \times 5 = 18$

(C) No, because $3 \times 5 = 15$

(D) No, because $3 + 5 = 8$

Find more at
ViewMath.com/CO-Grade6

ViewMath.com

19. *Write a real-world situation that can be described by $x \leq 15$.*

Your Answer:

20. *Look at the two triangles on the grid. Which triangle has a greater area?*

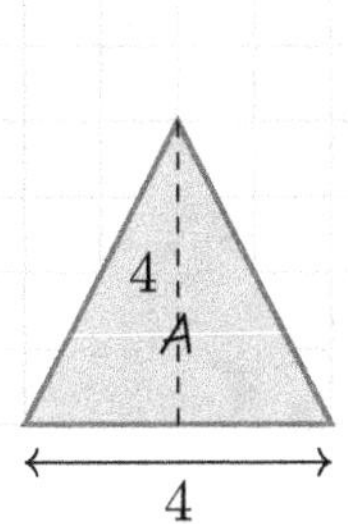

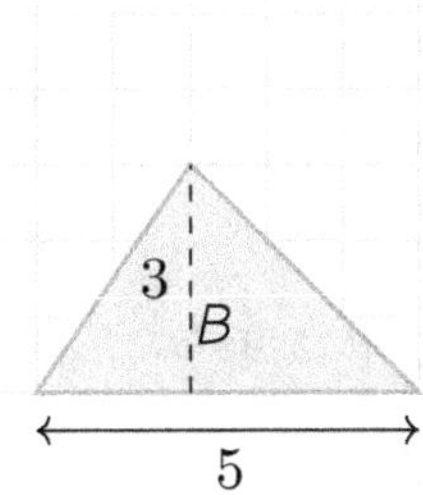

(A) Triangle A

(B) Triangle B

(C) They have the same area.

(D) Not enough information to tell.

21. *A parallelogram has base 4.5 m and height 6 m. What is the area?*

Your Answer:

22. *Which formula gives the volume of a rectangular prism?*

(A) $V = l + w + h$

(B) $V = 2lw + 2lh + 2wh$

(C) $V = l \times w \times h$

(D) $V = l \times w$

23. *Two points share the same y-coordinate. They form a:*

(A) Vertical segment

(B) Diagonal segment

(C) Horizontal segment

(D) There is not enough information to tell.

24. A rectangle has vertices $(2, -1)$, $(2, 5)$, $(9, 5)$, and $(9, -1)$. What is the area?

Your Answer:

25. A cube has a surface area of 216 cm^2. What is the edge length?

Your Answer:

26. Which of the following is a statistical question?

(A) What year was the school built?

(B) What is the name of our principal?

(C) How many minutes does each student exercise per day?

(D) How many continents are there?

27. Describe the shape of the data set: $1, 2, 3, 3, 4, 4, 4, 5, 5, 5, 5$.

Your Answer:

28. Which display would you use to show the heights of 200 students?

(A) Dot plot, because it shows individual values.

(B) Histogram, because it groups the large data set into intervals.

(C) Either works equally well.

(D) Neither works for numerical data.

29. A box plot shows: $min = 25$, $Q1 = 35$, $median = 50$, $Q3 = 65$, $max = 80$. Find the range and IQR.

Your Answer:

Find more at
ViewMath.com/CO-Grade6

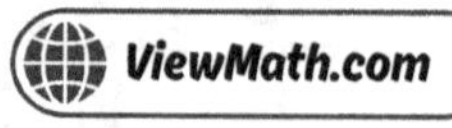

30. Two classes both have a mean of 75. Class A: MAD = 3. Class B: MAD = 9. A student from each class is picked at random. Whose score is more likely to be close to 75?

(A) The student from Class A

(B) The student from Class B

(C) Both equally likely

(D) Cannot determine

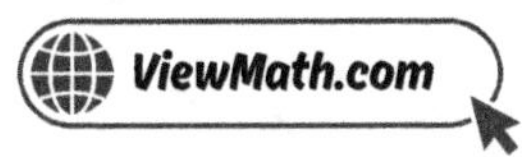

 End of Practice Test 8 

Great job finishing the test!

 **My Score**

I got _____________ out of 30 questions right.

Check your answers in the Answer Key at the back of the book.

Review any questions you missed. That's how we learn!

 **Check Your Score Online!**

Visit **ViewMath Academy** to enter your answers and see which topics you need to review. You can also explore lessons, take quizzes, track your scores, and save your progress!

viewmath.com/score/6.1.CO.23

Or go to viewmath.com/score and enter code: 6.1.CO.23

Practice Test 9

 30 Questions

---

✏️ Before You Start ✏️

- ✓ **Read each question carefully** before choosing your answer.
- ✓ **Show your work** on scratch paper when you need to.
- ✓ **Skip hard questions** and come back to them later.
- ✓ **Check your answers** when you're done.
- ✓ **Take your time** — there's no rush!

⭐ **You've Got This!** ⭐

Do your best and show what you know!

1. The ratio of red to blue to green beads is $1:3:2$. There are 18 beads total. How many blue beads are there?

(A) 3

(B) 6

(C) 9

(D) 12

2. A printer prints 120 pages in 4 minutes. What is the unit rate?

(A) 4 pages per minute

(B) 30 pages per minute

(C) 120 pages per minute

(D) 480 pages per minute

3. Look at this ratio table. What is the missing value?

Apples	Oranges
3	5
6	?

(A) 8

(B) 10

(C) 15

(D) 11

4. Does the point $(3, 9)$ belong on the graph of the ratio $1:3$?

(A) No, because $3 + 9 \neq 1 + 3$.

(B) Yes, because $1:3 = 3:9$.

(C) No, because $3 \times 3 \neq 1$.

(D) Yes, because $3 + 9 = 12$.

5. A batch of trail mix uses 4 cups of granola for every 3 cups of raisins. If you have 12 cups of raisins, how many cups of granola do you need?

(A) 9

(B) 12

(C) 16

(D) 15

Find more at
ViewMath.com/CO-Grade6

6. *A recipe needs 2.5 liters of water. How many milliliters is that?*

(A) 25

(B) 250

(C) 2,500

(D) 25,000

7. *Emma owes $500 on a credit card with 18% annual interest. She makes no payments for one year. What is her new balance?*

(A) $518

(B) $540

(C) $590

(D) $680

8. *The bar chart below shows how much two friends saved each month for 5 months. Each friend started with $0.*

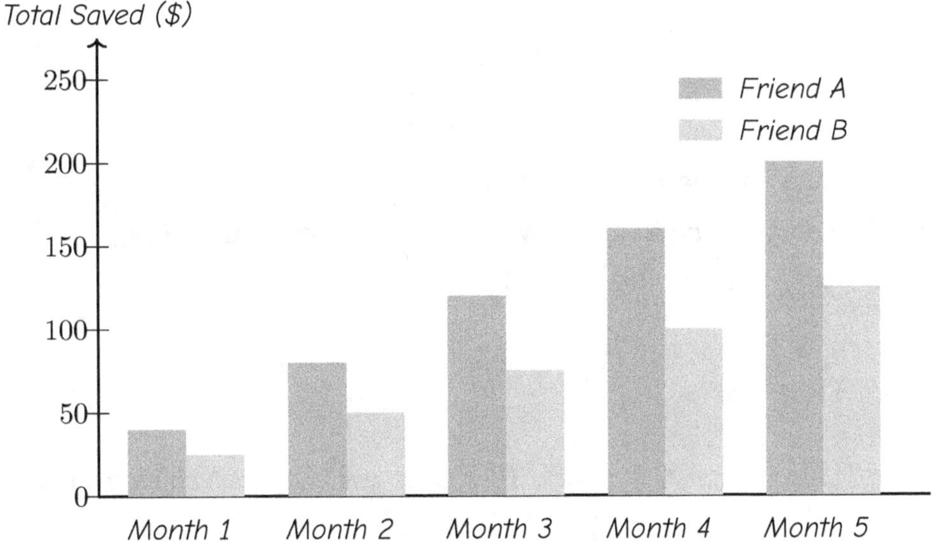

Friend A saves $40 per month and Friend B saves $25 per month.

Part A: After 5 months, how much more has Friend A saved than Friend B?

Part B: If Friend B deposits her 5-month total into an account earning 4% simple interest per year, how much interest will she earn in 1 year?

Your Answer

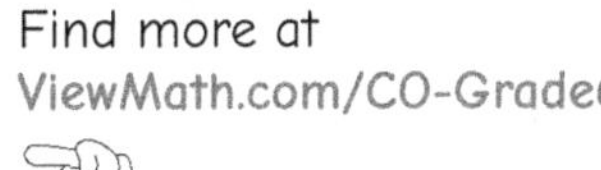

Find more at
ViewMath.com/CO-Grade6

9. *What is the value of $4(9+6)$?*

   (A) 40

   (B) 54

   (C) 60

   (D) 24

10. *The thermometer below shows the current temperature.*

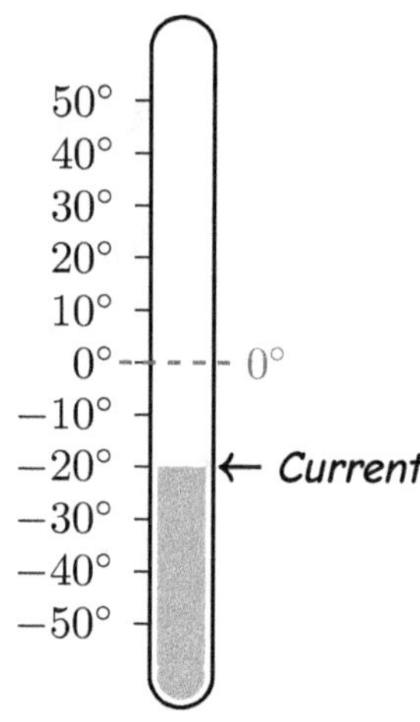

Part A: *What temperature does the thermometer show?*

Part B: *Is this temperature above or below zero? Write the temperature as an integer.*

Your Answer:

11. *The temperature on Monday was $-5°F$ and on Tuesday it was $-2°F$. Which statement is true?*

   (A) *Monday was warmer because $5 > 2$*

   (B) *Tuesday was warmer because $-2 > -5$*

   (C) *Both days had the same temperature*

   (D) *Tuesday was colder because -2 is negative*

12. *What is the distance between the points $(3, 4)$ and $(7, 4)$?*

   (A) -4

   (B) 4

   (C) 10

   (D) 11

Find more at
ViewMath.com/CO-Grade6

ViewMath.com

13. *Evaluate:* $48 \div (4 + 4)$

(A) 16   (B) 6

(C) 8   (D) 12

14. *Which expression represents "7 less than a number y"?*

(A) $7 - y$ (B) $y - 7$

(C) $7 + y$ (D) $7y$

15. *Evaluate $a^2 + b^2$ when $a = 3$ and $b = 4$.*

(A) 7   (B) 12

(C) 25   (D) 49

16. *Which expression is equivalent to $9w - 4w + 6 - 6$?*

(A) $5w$ (B) $5w + 12$

(C) $13w$ (D) 5

17. *A store sells notebooks for \$3 each and pens for \$1 each. Which expression gives the total cost for n notebooks and p pens?*

(A) $3 + n + 1 + p$ (B) $3n + p$

(C) $4np$ (D) $n + 3p$

18. Solve: $\dfrac{m}{4} = 7$

    (A) $m = 3$ (B) $m = 11$

    (C) $m = 28$ (D) $m = 47$

19. A parking lot allows no more than 200 cars. Which inequality describes the number of cars c?

    (A) $c > 200$ (B) $c \geq 200$

    (C) $c < 200$ (D) $c \leq 200$

20. A triangle has base $\frac{1}{2}$ ft and height $\frac{1}{4}$ ft. What is the area?

    (A) $\frac{1}{8}$ ft^2 (B) $\frac{3}{4}$ ft^2

    (C) $\frac{1}{16}$ ft^2 (D) $\frac{1}{4}$ ft^2

21. A composite figure is made of a rectangle (8 cm by 5 cm) attached to a triangle with base 8 cm and height 3 cm. What is the total area?

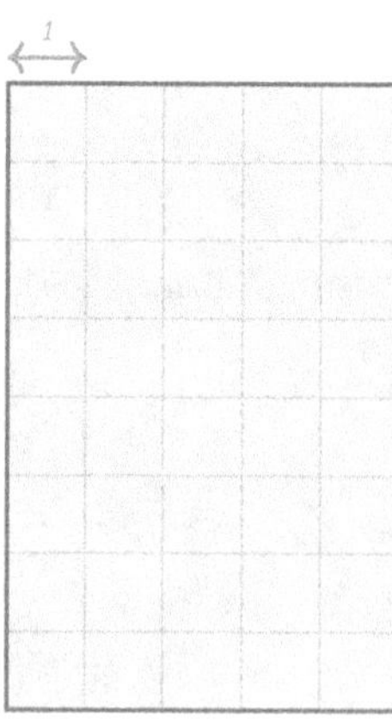

    (A) 52 cm^2 (B) 40 cm^2

    (C) 55 cm^2 (D) 64 cm^2

Find more at
ViewMath.com/CO-Grade6

22. What is the volume of a rectangular prism with length 5 cm, width 3 cm, and height 4 cm?

   (A) $12\ cm^3$ (B) $60\ cm^3$

   (C) $30\ cm^3$ (D) $94\ cm^3$

23. Points $(4, 1)$ and $(-2, 1)$ form one side of a rectangle. What is the length of that side?

   (A) 2 units   (B) 4 units

   (C) 6 units   (D) 3 units

24. A right triangle has vertices $(0, 0)$, $(8, 0)$, and $(0, 6)$. What is the area?

   (A) 48 square units   (B) 24 square units

   (C) 14 square units   (D) 28 square units

25. The net below folds into a rectangular prism. What is the surface area?

		4×2	
		4×3	
3×2		4×2	3×2
		4×3	

   (A) $52\ cm^2$ (B) $24\ cm^2$

   (C) $48\ cm^2$ (D) $96\ cm^2$

26. A survey question is: "How many minutes does it take you to get to school?" Which is true about the data collected?

(A) Every student will give the same answer.

(B) The answers will vary depending on each student's commute.

(C) The data cannot be displayed on a graph.

(D) The question cannot produce numerical data.

27. Look at the dot plot below.

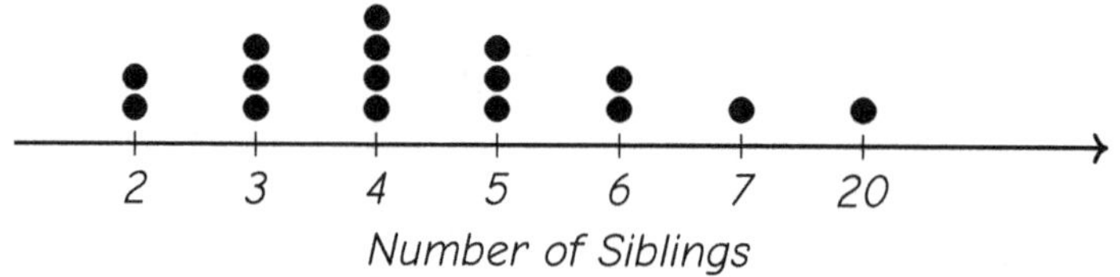

Which value is an outlier?

(A) 2

(B) 4

(C) 7

(D) 20

28. In a histogram, the bar for the interval 50–59 has a height of 8. What does this mean?

(A) The values range from 50 to 59.

(B) There are 8 data values between 50 and 59.

(C) The average of the values is 8.

(D) The interval is 8 units wide.

29. Data: $1, 3, 5, 7, 9, 11, 13, 15, 17, 19$. What is the IQR?

(A) 8

(B) 10

(C) 14

(D) 18

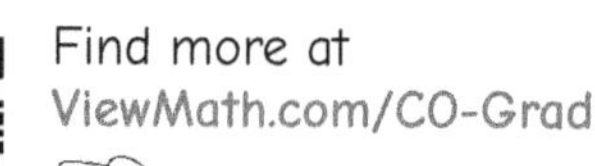

Find more at
ViewMath.com/CO-Grade6

30. *Two basketball players' points per game: Player A: 18, 20, 22, 24, 26. Player B: 10, 15, 22, 28, 35. Find the mean and range for each player and compare.*

Your Answer:

Find more at
ViewMath.com/CO-Grade6

 # End of Practice Test 9 

Great job finishing the test!

 My Score

I got _____________ out of 30 questions right.

Check your answers in the **Answer Key** at the back of the book.

Review any questions you missed. That's how we learn!

Check Your Score Online!

Visit **ViewMath Academy** to enter your answers and see which topics you need to review. You can also explore lessons, take quizzes, track your scores, and save your progress!

viewmath.com/score/6.1.CO.24

Or go to viewmath.com/score and enter code: 6.1.CO.24

Practice Test 10

30 Questions

✏ Before You Start ✏

- ✔ **Read each question carefully** before choosing your answer.
- ✔ **Show your work** on scratch paper when you need to.
- ✔ **Skip hard questions** and come back to them later.
- ✔ **Check your answers** when you're done.
- ✔ **Take your time** — there's no rush!

⭐ **You've Got This!** ⭐

Do your best and show what you know!

1. The tape diagram below shows the ratio of apple juice to orange juice in a drink.

Apple: ☐☐☐
Orange: ☐☐☐☐☐

If each part equals 4 ounces, how many total ounces of drink are there?

(A) 12

(B) 20

(C) 32

(D) 8

2. The table below shows the prices at two stores for packs of pencils.

	Number of Pencils	Price
Store A	10	$4.50
Store B	8	$3.20

Which store has the lower unit price per pencil?

(A) Store A at $0.45 per pencil

(B) Store B at $0.40 per pencil

(C) Store A at $0.40 per pencil

(D) They have the same unit price.

3. Which pair of ratios are equivalent?

(A) $2 : 3$ and $4 : 9$

(B) $2 : 3$ and $6 : 9$

(C) $2 : 3$ and $8 : 9$

(D) $2 : 3$ and $3 : 2$

Find more at
ViewMath.com/CO-Grade6

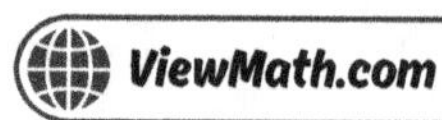

4. Which table matches a graph that passes through $(0,0)$, $(2,3)$, and $(4,6)$?

(A)

x	y
2	3
6	9

(B)

x	y
2	3
6	8

(C)

x	y
3	2
6	9

(D)

x	y
2	4
6	12

5. A fence uses posts in a ratio of 1 post for every 6 feet. How many posts are needed for a 42-foot fence?

(A) 6

(B) 7

(C) 8

(D) 36

6. The number line below shows a distance in centimeters.

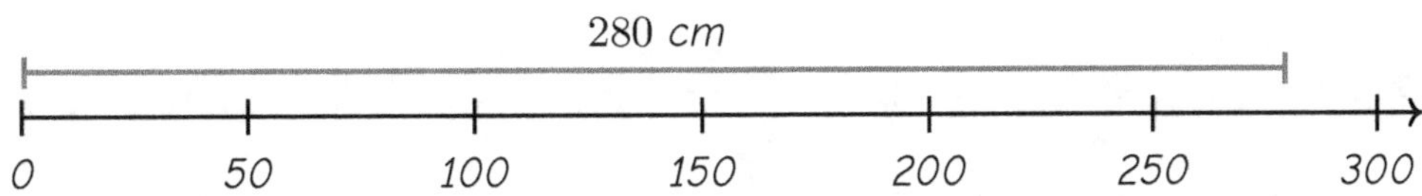

Part A: Convert 280 centimeters to meters.

Part B: Convert 280 centimeters to millimeters.

Part C: A doorway is 2 meters tall. Is 280 cm longer or shorter? By how much?

Your Answer:

7. Aiden earns \$350 per month walking dogs. He saves 30% of his income each month. How much does he save in one year?

Your Answer:

8. Emma wants to buy a $120 skateboard. She earns $40 per week and saves 50% of her income. How many weeks will it take her to save enough money?

> *Your Answer:*

9. What is the GCF that should be used to rewrite $36 + 8$ with the distributive property?

(A) 2   (B) 4

(C) 8   (D) 6

10. On a number line, where are negative numbers located?

(A) To the right of zero   (B) To the left of zero

(C) Above zero   (D) On top of zero

11. The number line below shows five points labeled J, K, L, M, and N.

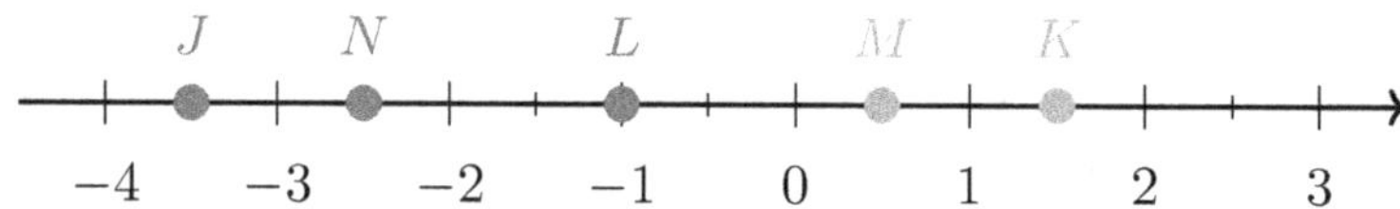

Part A: Write the value of each point.

Part B: List the five points in order from least to greatest value.

Part C: Which point has the greatest absolute value? Explain.

> *Your Answer:*

12. Two traffic lights are at $(-3, 5)$ and $(4, 5)$ along the same street on a city grid. How many blocks apart are they?

(A) 1

(B) 5

(C) 7

(D) 9

13. The area of each small square below is 1 square unit. Write the area using an exponent, then find the total area.

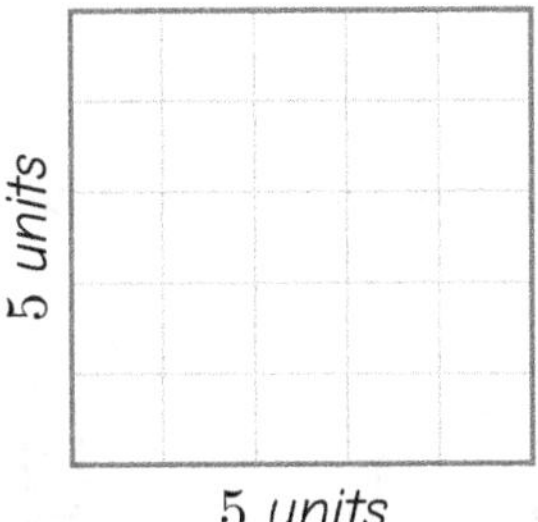

Your Answer

14. Which expression represents "6 more than half of a number n"?

(A) $6 + 2n$

(B) $\dfrac{n + 6}{2}$

(C) $\dfrac{n}{2} + 6$

(D) $n + 3$

15. Evaluate $10 - 2(k - 1)$ when $k = 4$.

(A) 2

(B) 4

(C) 6

(D) 14

16. *Simplify:* $9x + 2 + 3x - 7$

Your Answer:

17. *A pizza costs \$14 and is split equally among f friends. Which expression shows each friend's share?*

(A) $14f$

(B) $14 + f$

(C) $14 - f$

(D) $14 \div f$

18. *Look at the balance scale below. Each circle weighs x pounds. Each square weighs 1 pound. The scale is balanced. Write and solve the equation.*

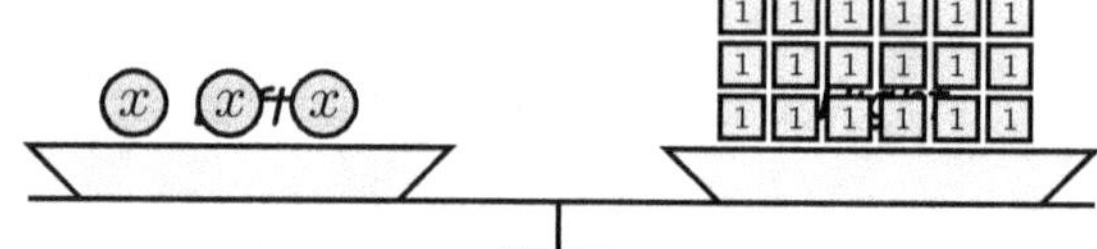

Your Answer:

19. *Which of the following is a solution to $n \leq 8$?*

(A) $n = 9$

(B) $n = 8.5$

(C) $n = 8$

(D) $n = 10$

20. *A triangle has an area of 24 ft^2 and a base of 8 ft. What is the height?*

(A) 3 ft

(B) 6 ft

(C) 16 ft

(D) 4 ft

Find more at
ViewMath.com/CO-Grade6

21. What is the area of a parallelogram with base 9 cm and height 5 cm?

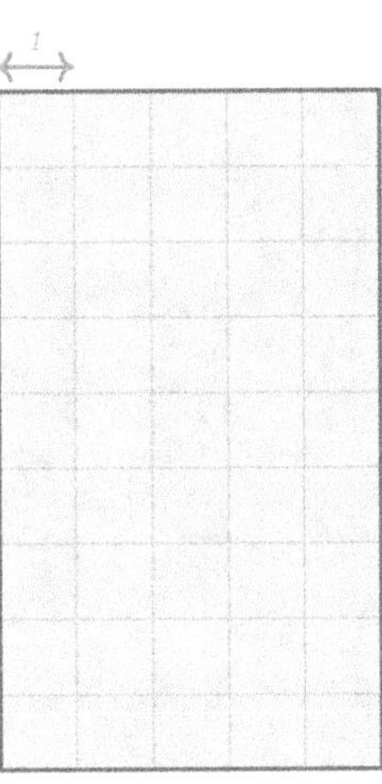

(A) $14\ cm^2$

(B) $45\ cm^2$

(C) $22.5\ cm^2$

(D) $28\ cm^2$

22. A toy box is 3 ft long, 2 ft wide, and 2 ft tall. What is the volume?

(A) $7\ ft^3$

(B) $12\ ft^3$

(C) $6\ ft^3$

(D) $24\ ft^3$

23. Three vertices of a rectangle are $(2, 3)$, $(2, -5)$, and $(9, 3)$. What is the fourth vertex?

Your Answer:

24. The L-shaped figure below is drawn on a grid. What is its area?

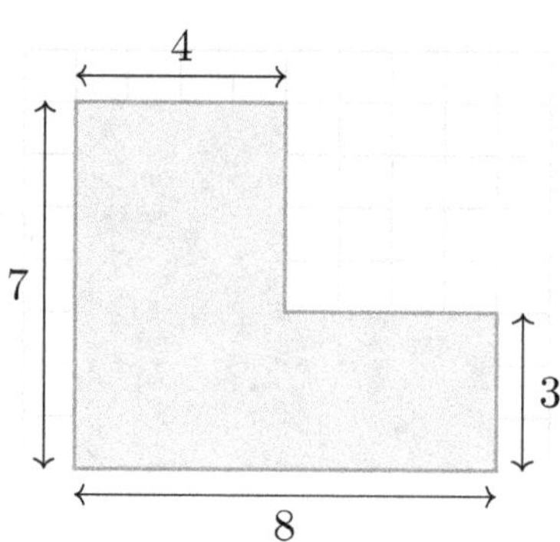

(A) 40 square units

(B) 32 square units

(C) 56 square units

(D) 24 square units

25. A cereal box is 20 cm tall, 30 cm wide, and 8 cm deep. What is the surface area?

(A) 1,520 cm^2

(B) 2,000 cm^2

(C) 1,840 cm^2

(D) 4,800 cm^2

26. Is the following question statistical or not statistical?

"How many text messages does each student send per day?"

Your Answer

27. Which term describes a data value that is much higher or lower than the rest of the data?

(A) Cluster

(B) Gap

(C) Outlier

(D) Peak

28. Look at the dot plot below showing the number of hours students studied for a test.

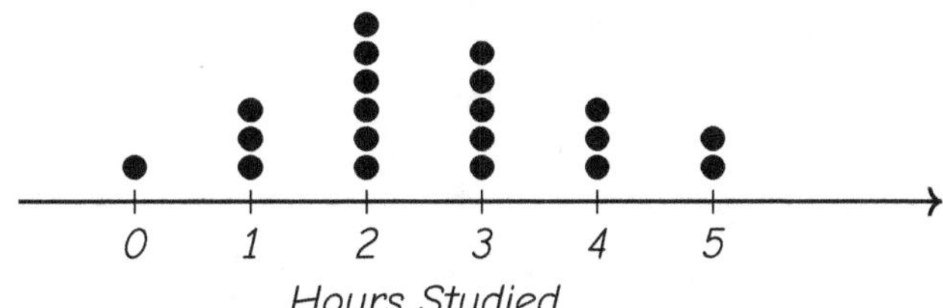

How many students studied 3 or more hours?

(A) 5

(B) 8

(C) 10

(D) 20

29. Data (in order): $8, 12, 14, 18, 22, 26, 30$. What is the five-number summary?

(A) $8, 12, 18, 26, 30$

(B) $8, 14, 18, 22, 30$

(C) $8, 12, 14, 26, 30$

(D) $8, 13, 18, 24, 30$

30. Data: $15, 18, 20, 22, 25$. The mean is 20 and the range is 10. Which is the best one-sentence summary?

(A) The data has 5 values.

(B) The typical value is about 20, and values span 10 units.

(C) The smallest value is 15.

(D) The data is not statistical.

⭐ *End of Practice Test 10* ⭐

Great job finishing the test!

 **My Score**

I got ______________ out of 30 questions right.

*Check your answers in the **Answer Key** at the back of the book.*

💡 *Review any questions you missed. That's how we learn!*

📊 Check Your Score Online!

Visit **ViewMath Academy** to enter your answers and see which topics you need to review. You can also explore lessons, take quizzes, track your scores, and save your progress!

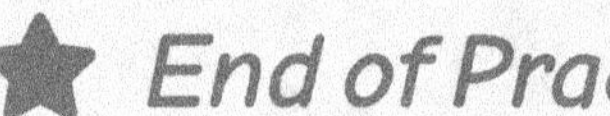

viewmath.com/score/6.1.CO.25

Or go to viewmath.com/score and enter code: 6.1.CO.25

Answer Key & Explanations

Answer Key

First try each test on your own, then check your work here.

☑ Practice Test 1 — Answer Key

 5
 C
 C
 C
 C
 8 full bottles
 C
 B

 $6(2+3+5)$
 -18
 C

 Part A: 7 blocks;   Part B: 6 blocks;   Part C: 4 blocks;   Part D: 17 blocks
 41
 C

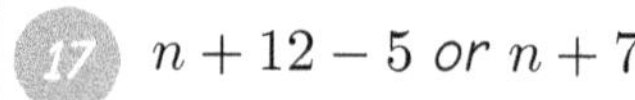 $23
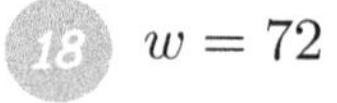 C
 $n + 12 - 5$ or $n + 7$
 $w = 72$
 C
 C
 C
 C

 B
 A
 C
 C
 C
 25
29 D
30 No.

💡 Time to Learn! 💡

Review the explanations below, **especially for the questions you missed**.

Understanding why each answer is correct builds stronger problem-solving skills.

Tip: Circle any questions you got wrong, then read their explanation carefully.

📖 Practice Test 1 — Detailed Explanations

1. *Each part* $= 20 \div 4 = 5$. *Vegetables* $= 1 \times 5 = 5$.

Find more at
ViewMath.com/CO-Grade6

2 From the graph, the truck travels 50 miles in 2 hours. Unit rate: $50 \div 2 = 25$ miles per hour.

3 From the double number line, 9 scoops aligns with 15 cups of water.

4 $20 \div 5 = 4$. The unit rate is 4 (units of y per 1 unit of x).

5 Scale: 2 cm = 15 km, so 1 cm = 7.5 km. For 5 cm: $5 \times 7.5 = 37.5$ km.

6 6 L = 6,000 mL. $6,000 \div 750 = 8$.

7 Amy: $0.20 \times \$80 = \$16$. Ben: $0.20 \times \$50 = \$10$. Cara: $0.20 \times \$100 = \$20$. Dan: $0.20 \times \$60 = \$12$. Cara saves the most.

8 $I = P \times r \times t = 500 \times 0.04 \times 2 = \$40$.

9 The GCF of 12, 18, and 30 is 6. Divide: $12 \div 6 = 2$, $18 \div 6 = 3$, $30 \div 6 = 5$. So $12 + 18 + 30 = 6(2 + 3 + 5) = 6 \times 10 = 60$.

10 Below zero means negative. 18 degrees below zero is written as -18.

11 Read the values: $A = -2\frac{3}{4}$, $B = -\frac{1}{4}$, $C = 1\frac{1}{2}$, $D = -1\frac{1}{2}$. The point farthest to the right on the number line has the greatest value. That is Point C at $1\frac{1}{2}$.

12 Part A: Home $(-4, 5)$ and Bakery $(3, 5)$ share $y = 5$. Distance $= |-4 - 3| = 7$. Part B: Bakery $(3, 5)$ and Gym $(3, -1)$ share $x = 3$. Distance $= |5 - (-1)| = 6$. Part C: Gym $(3, -1)$ and Pool $(3, -5)$ share $x = 3$. Distance $= |-1 - (-5)| = |-1 + 5| = 4$. Part D: Total $= 7 + 6 + 4 = 17$ blocks.

13 $5^2 = 25$ and $4^2 = 16$. Sum: $25 + 16 = 41$.

Find more at
ViewMath.com/CO-Grade6

14 *"Plus" means addition: $n + 8$.*

15 $5 + 3(6) = 5 + 18 = 23$ *dollars.*

16 *Distribute: $2 \times y + 2 \times 9 = 2y + 18$.*

17 *Start with n, add 12, subtract 5: $n + 12 - 5 = n + 7$.*

18 *Multiply by 8: $w = 9 \times 8 = 72$.*

19 $0 > -2$ *is true, so $x = 0$ is a solution. -2 is not greater than -2, and -3 and -2.5 are less than -2.*

20 *One sail: $\frac{1}{2} \times 3 \times 7 = 10.5\ m^2$. Two sails: $10.5 \times 2 = 21\ m^2$.*

21 *Both use $A = b \times h = 7 \times 4 = 28\ m^2$. They have equal area.*

22 *Volume measures 3-dimensional space, so it uses cubic units like cm^3, in^3, or ft^3.*

23 *Horizontal side: $|7 - 1| = 6$. Vertical side: $|2 - (-4)| = 6$. Both sides are 6 units, so it is a square. The side length is 6.*

24 *Rectangle: $5 \times 4 = 20$. Triangle: $\frac{1}{2} \times 5 \times 3 = 7.5$. Total: $20 + 7.5 = 27.5$ square units.*

25 *2 triangular bases $+3$ rectangular faces $= 5$ faces total.*

26 *If you ask 10 people how long they spend eating lunch, the answers vary. Following the flowchart, varying answers lead to "Statistical."*

Find more at
ViewMath.com/CO-Grade6

27. The value 6 appears three times (the most), and the data clusters around 5–7. The peak is at 6.

28. $5 + 10 + 8 + 2 = 25$ students.

29. 9 values. Median is the 5th value (40). Upper half: $45, 50, 55, 60.$ $Q3 = (50 + 55) \div 2 = 52.5$.

30. Range only measures spread, not center. A smaller range means more consistency but doesn't say anything about whether scores were higher or lower. You need to compare centers too.

✅ Practice Test 2 — Answer Key

1. C

2. Part A: Lily earns $12 per hour, Noah earns $13 per hour.   Part B: Noah earns $1 more per hour.

3. Part A: 2 : 1;   Part B: 4 cups of sugar

4. C

5. B

6. C

7. C

8. 27.5%

9. B

10. Answers vary. Example: an elevation of 50 feet below sea level, or a debt of $50.

11. A

12. 14 units

13. $3^3 = 27$ unit cubes

14. $12h + 20$

15. C

16. A

17. B

18. C

19. $w < 50$

20. D

21. A

22. $500\ m^3$

23. C

24. 12 square units

25. A

26. Neither is statistical because the answers do not vary.

27. C

28. C

29. A

30. B

💡 Time to Learn! 💡

Review the explanations below, **especially for the questions you missed.**

Understanding why each answer is correct builds stronger problem-solving skills.

Tip: Circle any questions you got wrong, then read their explanation carefully.

Find more at
ViewMath.com/CO-Grade6

📖 Practice Test 2 — Detailed Explanations

1. Girls have 6 parts $= 24$, so each part $= 24 \div 6 = 4$. Boys have 4 parts $= 4 \times 4 = 16$.

2. Lily: $\$60 \div 5 = \$12/hr$. Noah: $\$52 \div 4 = \$13/hr$. Noah earns $\$13 - \$12 = \$1$ more per hour.

3. Part A: From the graph, $(2, 1)$ shows 2 cups of flour for every 1 cup of sugar. Part B: Following the pattern, $(8, 4)$, so 4 cups of sugar.

4. At $(2, 5)$: $x : y = 2 : 5$. Check: $(4, 10)$ simplifies to $2 : 5$ as well.

5. $360 \div 12 = 30$ minutes.

6. $1\ km = 1,000\ m = 100,000\ cm$. $10 \times 100,000 = 1,000,000\ cm$.

7. 25% of $\$120 = 0.25 \times 120 = \$30$.

8. Total expenses $= \$960 + \$320 + \$480 + \$240 + \$320 = \$2,320$. Savings $= \$3,200 - \$2,320 = \$880$. Rate $= \dfrac{880}{3,200} = 0.275 = 27.5\%$.

9. The GCF of 20 and 15 is 5. Divide: $20 \div 5 = 4$ and $15 \div 5 = 3$. So $20 + 15 = 5(4 + 3)$. Choice A is a common error: writing $15 \div 5 = 5$ instead of 3.

10. -50 represents being 50 units below a starting point. Any context where "below" or "less than zero" applies is correct.

11. From least to greatest: $-3.5 < -3.1 < -0.8 < 1.2 < 2.5$. Choice D shows greatest to least. Choices B and C have the negative numbers in the wrong order.

Find more at
ViewMath.com/CO-Grade6

12. A to B: same $y = 2$, distance $= |-4-3| = 7$. B to C: same $x = 3$, distance $= |2-(-5)| = 7$. Total distance $= 7 + 7 = 14$ units.

13. Each edge is 3 units. The cube needs $3 \times 3 \times 3 = 3^3 = 27$ unit cubes.

14. Hourly pay is $12h$. Plus the bonus: $12h + 20$.

15. $(10 + 6) \div 2 = 16 \div 2 = 8$.

16. $4(n + 2) = 4n + 8$ by the distributive property. They are equivalent for every value of n.

17. A square has 4 equal sides, so its perimeter is 4 times the side length: $P = 4s$.

18. Multiply by 6: $t = 5 \times 6 = 30$.

19. "Less than 50" means strictly under 50: $w < 50$.

20. $A = \frac{1}{2} \times 14 \times 8 = 56\ m^2$.

21. $A = 8 \times 3.5 = 28\ m^2$.

22. $V = 25 \times 10 \times 2 = 500\ m^3$.

23. Vertical segment: $|7 - (-1)| = |8| = 8$ units.

24. Rectangle area $= 6 \times 4 = 24$. Triangle area $= \frac{1}{2} \times 6 \times 4 = 12$. Remaining $= 24 - 12 = 12$ square units.

Find more at
ViewMath.com/CO-Grade6

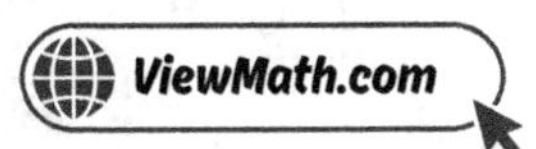

25. Net A is a valid cross-shaped cube net. Net B has two squares on the same side, which causes overlap. Net C forms a 2×3 block, which doesn't fold into a cube.

26. Every student gave the same answer for each question (3 for Q1, 12 for Q2). Since the data does not vary, neither question is statistical.

27. Range $= 18 - 12 = 6$. The range tells you how spread out the data is.

28. There are no data values between 25 and 30, creating a gap in the data.

29. Class A's box (IQR) is narrower than Class B's. A narrower box means the middle 50% of scores are closer together — more consistent.

30. Class X ranges from 25 to 40 (range $= 15$). Class Y ranges from 15 to 50 (range $= 35$). Class Y is much more spread out.

✓ Practice Test 3 — Answer Key

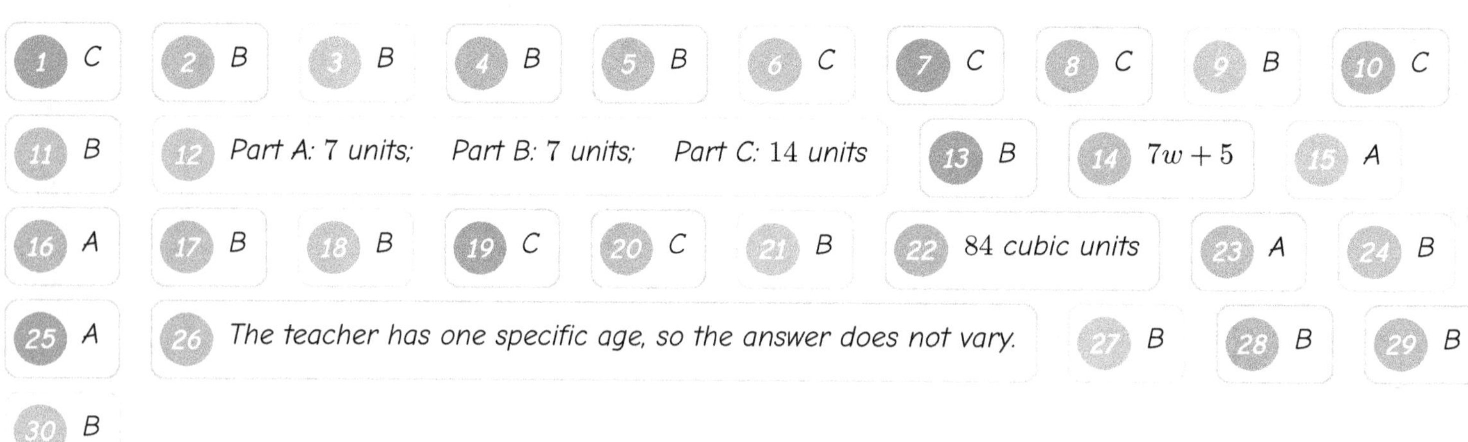

1 C	2 B	3 B	4 B	5 B	6 C	7 C	8 C	9 B	10 C
11 B	12 Part A: 7 units; Part B: 7 units; Part C: 14 units				13 B	14 $7w + 5$	15 A		
16 A	17 B	18 B	19 C	20 C	21 B	22 84 cubic units	23 A	24 B	
25 A	26 The teacher has one specific age, so the answer does not vary.					27 B	28 B	29 B	
30 B									

Find more at
ViewMath.com/CO-Grade6

💡 Time to Learn! 💡

*Review the explanations below, **especially for the questions you missed**.*

Understanding why each answer is correct builds stronger problem-solving skills.

Tip: *Circle any questions you got wrong, then read their explanation carefully.*

📖 Practice Test 3 — Detailed Explanations

1. The question asks flour to eggs. Flour $= 3$, eggs $= 2$. The ratio is $3 : 2$.

2. $\$9 \div 12 = \$0.75$ per muffin.

3. $7 \times 3 = 21$ bananas, so $\$2 \times 3 = \$6$.

4. The ratio is $6 : 2 = 3 : 1$. When $x = 15$: $y = 15 \div 3 = 5$.

5. Unit rate: $\$24 \div 3 = \$8$ per shirt. 10 shirts: $10 \times \$8 = \$80$.

6. $3 \times 5{,}280 = 15{,}840$ feet.

7. 30% of $\$4{,}000 = 0.30 \times 4{,}000 = \$1{,}200$.

8. Savings $=$ Income $-$ Expenses $= \$600 - \$450 = \$150$.

9. The GCF of 12 and 28 is 4. Divide: $12 \div 4 = 3$ and $28 \div 4 = 7$. So $12 + 28 = 4(3 + 7)$. Choice C forgets to divide 28 by the GCF.

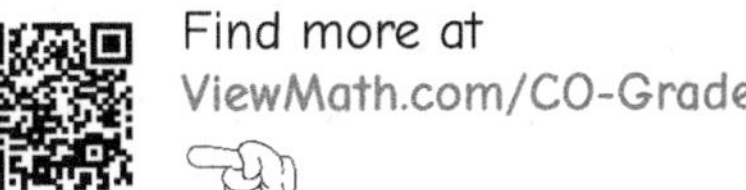
Find more at
ViewMath.com/CO-Grade6

10. $5°F$ is positive (above zero) and $-8°F$ is negative (below zero).

11. Both fractions have the same denominator. Since $-\frac{7}{8}$ is farther from zero (to the left of $-\frac{3}{8}$ on a number line), it is the smaller number. $-\frac{7}{8} < -\frac{3}{8}$.

12. Part A: Same $y = 3$, distance $= |-5 - 2| = |-7| = 7$. Part B: Same $x = 2$, distance $= |3 - (-4)| = |7| = 7$. Part C: Total $= 7 + 7 = 14$ units.

13. Any number raised to the power of 0 equals 1.

14. The product of 7 and w is $7w$. Increased by 5 gives $7w + 5$.

15. $2(3(5) - 1) = 2(15 - 1) = 2(14) = 28$.

16. Like terms: $5p + 2p + 4p = 11p$. Constants: $3 - 1 = 2$. Result: $11p + 2$.

17. When $t = 1$: $12(1) + 4 = 16$. When $t = 2$: $12(2) + 4 = 28$. Each shirt costs \$12 and shipping is \$4.

18. Add 11: $p = 25 + 11 = 36$.

19. "At most 21" means 21 or fewer. $d \le 21$.

20. $A = \frac{1}{2} \times 5 \times 12 = 30$ in^2.

21. $A = \frac{1}{2}(9 + 15)(6) = \frac{1}{2}(24)(6) = 72$ in^2.

22. Bottom prism: $8 \times 3 \times 2 = 48$. Top prism: $3 \times 3 \times 4 = 36$. Total: $48 + 36 = 84$ cubic units.

Find more at
ViewMath.com/CO-Grade6

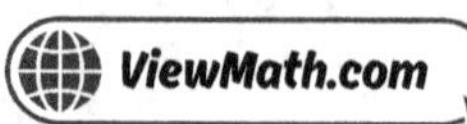

23 From $(-2, -2)$, moving 5 right gives $x = 3$, and 5 up gives $y = 3$. The opposite vertex is $(3, 3)$.

24 Length $= |3 - (-5)| = 8$. Width $= |4 - (-1)| = 5$. Area $= 8 \times 5 = 40$. Jake is correct.

25 $SA = 2(5)(3) + 2(5)(4) + 2(3)(4) = 30 + 40 + 24 = 94$ cm^2. Yes, it is correct.

26 A statistical question expects data that varies. The teacher's age is a single fixed number.

27 Data set A range: $24 - 20 = 4$. Data set B range: $34 - 10 = 24$. Data set B is much more spread out.

28 Total $= 2 + 5 + 4 + 3 + 1 = 15$ students.

29 The box represents Q1 to Q3, which contains the middle 50% of the data.

30 Little overlap means most values in one group are different from the values in the other group, which signals a clear difference between the groups.

☑ Practice Test 4 — Answer Key

1 B   2 3.5 hours   3 C   4 Line A. It has a rate of 6 per 1, which is higher than 4 per 1.

5 B   6 D   7 B   8 $90   9 B   10 C   11 C

12 8 units. Subtract the x-coordinates: $5 - (-3) = 8$. Take the absolute value: $|8| = 8$. The absolute value ensures distance

13 C   14 $\frac{t}{4} - 1$ 15 C   16 B   17 B   18 A   19 D   20 C   21 B   22 A

23 B   24 64 square units   25 B   26 B   27 C   28 C

Find more at
ViewMath.com/CO-Grade6

29 *Store B has higher typical sales (median ≈ $275 vs. $250). Store A has more predictable sales (IQR ≈ $100 vs.*

30 *B*

💡 *Time to Learn!* 💡

*Review the explanations below, **especially for the questions you missed**.*

Understanding why each answer is correct builds stronger problem-solving skills.

***Tip:** Circle any questions you got wrong, then read their explanation carefully.*

📖 **Practice Test 4 — Detailed Explanations**

1 *"3 pencils for every 1 eraser" means pencils to erasers is $3 : 1$.*

2 $49 \div 14 = 3.5$ *hours.*

3 *The ratio is $4 : 10 = 2 : 5$. Row 3 should be $12 : 30$ (since $4 \times 3 = 12$ and $10 \times 3 = 30$), but it shows $12 : 25$.*

4 *Line A goes up 6 for every 1 unit right; Line B goes up 4. Higher rise per unit means a faster rate.*

5 *Unit rate: $180 \div 3 = 60$ mph. In 7 hours: $60 \times 7 = 420$ miles.*

6 $3 \times 4 = 12$ *cups.*

7 *Credit interest owed: $\$150 \times 0.10 = \$15$. Savings interest earned: $\$150 \times 0.02 = \$3$. Difference: $\$15 - \$3 = \$12$.*

8 $I = P \times r \times t = 600 \times 0.05 \times 3 = \$90.$

Find more at
ViewMath.com/CO-Grade6

9   The GCF of 16 and 40 is 8. Divide: $16 \div 8 = 2$ and $40 \div 8 = 5$. So $16 + 40 = 8(2 + 5)$. Choices A and C use common factors that are not the greatest.

10   By definition, zero is neither positive nor negative. It is the boundary between positive and negative numbers.

11   Convert to decimals: -1.5, $\dfrac{3}{2} = 1.5$, $-\dfrac{7}{4} = -1.75$, 0. The least value is -1.75, which is $-\dfrac{7}{4}$. A common error is choosing -1.5 without checking $-\dfrac{7}{4}$.

12   Since both points share $y = -1$, the distance is $|5 - (-3)| = |5 + 3| = |8| = 8$. Absolute value removes any negative sign, because distance measures how far apart two points are and cannot be negative.

13 $3^4 = 3 \times 3 \times 3 \times 3 = 81$. Each value is 3 times the previous one: $27 \times 3 = 81$.

14   Divide t by 4 to get $\frac{t}{4}$, then subtract 1: $\frac{t}{4} - 1$.

15 $3^2 + 3 = 9 + 3 = 12$.

16   Group 1 has $3x + 1$. Group 2 has $x + 3$. Total: $3x + 1 + x + 3 = 4x + 4$. Choices B and D both represent the same value; B is simplified.

17 $d = 60t$ means distance equals 60 miles per hour times the number of hours.

18   Subtract 3.5: $x = 10 - 3.5 = 6.5$.

19 $2.5 \geq 3$ is false. $2.5 < 3$, so it is not a solution.

20 $A = \frac{1}{2} \times 20 \times 7 = 70$ in^2.

21   Area uses the height, not the slant side. $A = 12 \times 6 = 72$ m^2.

Find more at
ViewMath.com/CO-Grade6

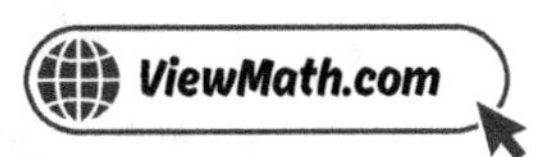

22 $V = \frac{1}{2} \times \frac{1}{2} \times \frac{1}{2} = \frac{1}{8} \ m^3$.

23 $A(1, -2)$ and $B(7, -2)$ have the same y-coordinate. Distance $= |7 - 1| = 6$ units.

24 Bottom rectangle: $10 \times 4 = 40$. Top-left rectangle: $6 \times (8 - 4) = 6 \times 4 = 24$. Total: $40 + 24 = 64$ square units.

25 A net is a flat pattern that, when folded, forms a 3D shape. Each section of the net becomes a face.

26 Different eagles have different wingspans. The data would vary from eagle to eagle.

27 Most data values are between 8 and 12, clustering around 10. The 25 is an outlier. A typical time is about 10 minutes.

28 Total $= 3 + 7 + 10 + 5 = 25$.

29 Store A: median $\approx \$250$, IQR $\approx 300 - 200 = \$100$. Store B: median $\approx \$275$, IQR $\approx 350 - 225 = \$125$. Store B sells more on a typical day, but Store A's sales are more consistent (smaller IQR and range).

30 Runner B has a lower mean time ($24 < 25$), so faster. Runner A has a smaller MAD ($1 < 5$), so more consistent.

✅ Practice Test 5 — Answer Key

1 10   **2** C   **3** C   **4** B   **5** C   **6** 150 minutes; 9,000 seconds   **7** D   **8** C

9 A   **10** 0   **11** $-1.25, \ -\frac{2}{3}, \ -0.1, \ 0.5, \ \frac{3}{4}$ **12** C   **13** 4   **14** D   **15** A

16 Yes. $5(3 + 2) = 25$ and $5(3) + 10 = 25$ **17** C   **18** B   **19** C   **20** B   **21** 7 cm

Find more at
ViewMath.com/CO-Grade6

 A    A    C    C    B    The data clusters around 5–7. The peak is at 6.

 B   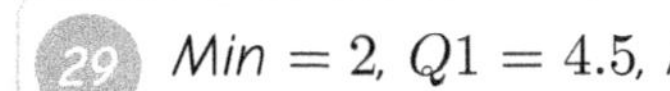 Min = 2, Q1 = 4.5, Median = 8, Q3 = 11, Max = 15. IQR = 6.5.    B

💡 **Time to Learn!** 💡

Review the explanations below, **especially for the questions you missed**.

Understanding why each answer is correct builds stronger problem-solving skills.

Tip: Circle any questions you got wrong, then read their explanation carefully.

📖 Practice Test 5 — Detailed Explanations

1 Total parts $= 3 + 2 = 5$. Each part $= 25 \div 5 = 5$. Swimming $= 2 \times 5 = 10$.

2 $240 \div 8 = 30$ miles per gallon.

3 $4 \times 2.5 = 10$ loaves, so $6 \times 2.5 = 15$ cups. Alternatively, unit rate: $6 \div 4 = 1.5$ cups per loaf, then $1.5 \times 10 = 15$.

4 Ratio: $1 : 4$. When $x = 3$: $y = 3 \times 4 = 12$.

5 $8 \times 5 = 40$ km.

6 $2 \times 60 + 30 = 150$ minutes. $150 \times 60 = 9{,}000$ seconds.

7 A debit card takes money directly from your bank account, unlike a credit card which borrows money.

Find more at
ViewMath.com/CO-Grade6

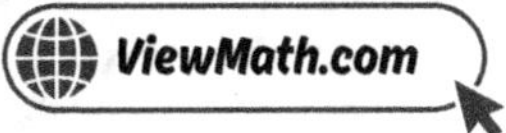

8. Savings percent $= 100\% - 35\% - 10\% - 20\% - 15\% = 20\%$. Before: $0.20 \times \$1{,}000 = \$200$. After: $0.20 \times \$1{,}200 = \$240$. Difference $= \$240 - \$200 = \$40$.

9. $5 \times 46 = 5(40 + 6) = 5 \times 40 + 5 \times 6 = 200 + 30 = 230$.

10. He gained $20 and then spent $20, so the overall change is $0. He is back to where he started.

11. Convert to decimals: $-\dfrac{2}{3} \approx -0.667$, 0.5, -1.25, $\dfrac{3}{4} = 0.75$, -0.1. Ordering: $-1.25 < -0.667 < -0.1 < 0.5 < 0.75$.

12. The points share the same x-coordinate, so the distance is $|-6 - (-1)| = |-6 + 1| = |-5| = 5$. Choice B forgets the absolute value. Choice D adds the absolute values $6 + 1$ instead of subtracting.

13. Parentheses: $12 - 4 = 8$. Exponent: $8^2 = 64$. Divide: $64 \div 16 = 4$.

14. Giving away 6 means subtracting: $x - 6$.

15. $20 - 3(4) = 20 - 12 = 8$.

16. Both give 25 when $x = 3$, and the distributive property confirms they are equivalent for all values.

17. s represents the side length of any square. It can be any positive number.

18. Add 8 to both sides: $y = 14 + 8 = 22$.

19. You can drive at 65 or below, so ≤ 65.

20. $P = \frac{1}{2} \times 10 \times 6 = 30$. $Q = \frac{1}{2} \times 10 \times 8 = 40$. Difference $= 40 - 30 = 10\ cm^2$.

Find more at
ViewMath.com/CO-Grade6

21 $84 = \frac{1}{2}(10 + 14)h = \frac{1}{2}(24)h = 12h.$ So $h = 84 \div 12 = 7$ cm.

22   Base area $= 10 \times 4 = 40.$ Height $= 120 \div 40 = 3$ cm.

23   From $(0, 0)$ with length 10 along the x-axis and width 4 along the y-axis, the opposite vertex is $(10, 4)$.

24   Length $= 6$, width $= 4.$ Area $= 6 \times 4 = 24$ square units.

25   Surface area measures flat space, so it uses square units like cm^2 or in^2.

26   There are always 3 feet in one yard. This is not a statistical question because the answer does not vary.

27   The value 6 appears three times (the most). Values between 5 and 7 contain most of the data.

28   Most data clusters between 40 and 50. The values 10 and 80 are far from the cluster and may be outliers.

29   9 values, already ordered. Median is the 5th value: 8. Lower half: $2, 4, 5, 7.$ $Q1 = (4 + 5) \div 2 = 4.5.$ Upper half: $9, 10, 12, 15.$ $Q3 = (10 + 12) \div 2 = 11.$ $IQR = 11 - 4.5 = 6.5.$

30   The range is 12 and the data is evenly spaced. Whether "very spread out" is accurate depends on context, but the data is quite orderly and not wildly spread.

✅ Practice Test 6 — Answer Key

1   Bananas to yogurt: $3 : 4;$ Yogurt to bananas: $4 : 3$  **2**   B    **3**   B   **4**   A   **5**   B   **6**   B

7   B   **8**   B   **9**   C   **10**   C   **11**   B   **12**   D   **13**   8   **14**   D   **15**   C   **16**   B

Find more at
ViewMath.com/CO-Grade6

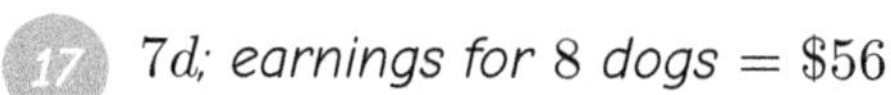 7d; earnings for 8 dogs = $56    $45    C   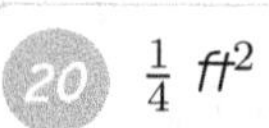 $\frac{1}{4}$ ft²    B    4 cm    B

 B    B    C    B    C    B    C

💡 Time to Learn! 💡

*Review the explanations below, **especially for the questions you missed**.*

Understanding why each answer is correct builds stronger problem-solving skills.

Tip: *Circle any questions you got wrong, then read their explanation carefully.*

📖 Practice Test 6 — Detailed Explanations

1. "3 bananas for every 4 cups of yogurt" gives 3 : 4. Flip the order for yogurt to bananas: 4 : 3.

2. Brand X: $8.75 \div 5 = \$1.75$ each. Brand Y: $4.50 \div 3 = \$1.50$ each. Brand Y is cheaper.

3. $2 \times 8 = 16$ students, so $3 \times 8 = 24$ pencils.

4. Hours go on the x-axis, miles on the y-axis: $(1, 30)$, $(2, 60)$, $(3, 90)$.

5. "How much for one" is a unit rate problem. Divide to find the amount per 1 unit.

6. 1 kg $= 1,000$ g. $7,000 \div 1,000 = 7$ kg.

7. $I = 600 \times 0.05 \times 3 = \$90$. Total $= \$600 + \$90 = \$690$.

8. Weekly savings $= 0.25 \times \$120 = \$30$. Total after 6 weeks $= \$30 \times 6 = \$180$. Interest $= 180 \times 0.03 \times 1 = \$5.40$.

Find more at
ViewMath.com/CO-Grade6

9 The GCF of 24 and 18 is 6. Divide each term: $24 \div 6 = 4$ and $18 \div 6 = 3$. So $24 + 18 = 6(4 + 3)$. Choices A and B use common factors that are not the GCF. Choice D has the wrong quotient for $18 \div 6$.

10 Zero in a bank account means no savings and no debt. It is the boundary between having money (positive) and owing money (negative).

11 Check each pair. A: $-\dfrac{1}{2} = -0.5$ and $-\dfrac{3}{4} = -0.75$; since $-0.5 > -0.75$, this is greatest to least (wrong). B: $-0.9 < -0.6$, which is least to greatest — correct! C: $-2.1 > -2.8$, so this is greatest to least (wrong). D: $\dfrac{1}{4} > -\dfrac{1}{4}$, so this is greatest to least (wrong).

12 Point A is at $(-3, 2)$ and Point B is at $(4, 2)$. They share $y = 2$, so the distance is $|-3 - 4| = |-7| = 7$ units. Choice A uses the y-value. Choice C uses $|-3 + 2|$ or a similar mix.

13 Parentheses: $3 + 2 = 5$. Exponent: $2^2 = 4$. Divide: $60 \div 5 = 12$. Subtract: $12 - 4 = 8$.

14 "Divide 20 by p" puts 20 first: $20 \div p$.

15 $15 \div 3 + 8 = 5 + 8 = 13$.

16 $10a - 2a = 8a$. The constant stays: $8a + 7$.

17 $7(8) = 56$ dollars.

18 $\dfrac{g}{5} = 9$, so $g = 9 \times 5 = 45$ dollars.

19 "Fewer than 20" means less than 20: $y < 20$.

20 $A = \frac{1}{2} \times \frac{3}{4} \times \frac{2}{3} = \frac{1}{2} \times \frac{6}{12} = \frac{1}{2} \times \frac{1}{2} = \frac{1}{4}$ ft².

21 When both bases are equal ($b_1 = b_2 = 4$), the trapezoid is actually a parallelogram. Area $= \frac{1}{2}(4+4)(6) = 24 = 4 \times 6$.

22 Base area $= 9 \times 5 = 45$. Height $= 180 \div 45 = 4$ cm.

23 $(5, -1)$ and $(5, 4)$ share the same x-coordinate, so the segment is vertical.

24 Decompose irregular shapes into simpler shapes (rectangles and triangles) whose areas you can compute, then add them.

25 One block: $2(3)(2) + 2(3)(4) + 2(2)(4) = 12 + 24 + 16 = 52$ ft^2. Two blocks: $52 \times 2 = 104$ ft^2.

26 Rainfall amounts change each month, so the data you collect to answer this question would vary.

27 All values are the same (50). The data is perfectly symmetric, and the range is $50 - 50 = 0$, so there is no spread.

28 Range $= 78 - 70 = 8$.

29 $IQR = Q3 - Q1 = 40 - 20 = 20$.

30 A complete summary uses center (mean or median) to describe what is typical and spread (range, IQR, or MAD) to describe consistency.

☑ Practice Test 7 — Answer Key

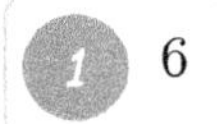 6
 B
 16 and 20
 C
 $150 and $200
 C
 $15
 C

Find more at
ViewMath.com/CO-Grade6

 B　 C　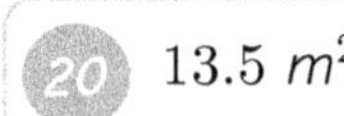 $0,\ -0.4,\ -\dfrac{1}{2},\ -0.55,\ -\dfrac{3}{5}$　 D　 B　 B　 45　 A

 B　 D　 B　 $13.5\ m^2$　 D　22 D　23 12 units　24 B

25 $268\ m^2$　26 C　27 19 students. The data has a peak in the 20–29 minute interval.　28 C

29 B　30 B

💡 Time to Learn! 💡

Review the explanations below, **especially for the questions you missed**.

Understanding why each answer is correct builds stronger problem-solving skills.

Tip: Circle any questions you got wrong, then read their explanation carefully.

📖 Practice Test 7 — Detailed Explanations

1. $21 \div 7 = 3$, so multiply both by 3: iced tea $= 2 \times 3 = 6$.

2. $150 \div 5 = 30$ gallons per hour.

3. Row 2: $5 \times 2 = 10$, so $8 \times 2 = 16$. Row 3: $8 \times 4 = 32$, so $5 \times 4 = 20$.

4. $5 : 15$ simplifies to $1 : 3$.

5. Total parts $= 7$. Each part $= \$350 \div 7 = \$50$. Family 1: $3 \times \$50 = \$150$. Family 2: $4 \times \$50 = \$200$.

6. 1 foot $= 12$ inches. $5 \times 12 = 60$ inches.

Find more at
ViewMath.com/CO-Grade6

7. Used: $40\% + 25\% + 10\% = 75\%$. Left: $100\% - 75\% = 25\%$. Amount: $0.25 \times \$60 = \$15$.

8. Savings rate $= \dfrac{\$50}{\$250} = 0.20 = 20\%$.

9. By the distributive property, $a(b + c) = a \times b + a \times c$. So $3(7 + 4) = 3 \times 7 + 3 \times 4 = 21 + 12 = 33$.

10. Above sea level is represented by a positive integer. 1,200 feet above sea level is $+1,200$ or simply $1,200$.

11. Convert to decimals: $-\dfrac{1}{2} = -0.5$, -0.4, $-\dfrac{3}{5} = -0.6$, 0, -0.55. From greatest to least: $0 > -0.4 > -0.5 > -0.55 > -0.6$.

12. The points share $y = 1$, so the distance is $|-4 - 2| = |-6| = 6$. Choice A forgets the absolute value. Choice B computes $-4 + 2$ without absolute value. Choice C computes $|-4 + 2|$.

13. Exponent first: $3^2 = 9$. Then multiply: $2 \times 9 = 18$.

14. "r squared" is r^2. "Minus 10" gives $r^2 - 10$.

15. $8(6) - 3 = 48 - 3 = 45$.

16. $7n$ and $4n$ are like terms. Add coefficients: $7 + 4 = 11$, so $7n + 4n = 11n$.

17. The one-time fee is $25. Monthly cost is $10m$. Total: $25 + 10m$.

18. Division is undone by multiplication. Multiply both sides by 3: $n = 12 \times 3 = 36$.

19. "Over 13" means greater than 13: $a > 13$.

Find more at
ViewMath.com/CO-Grade6

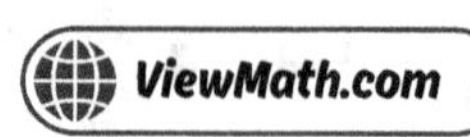

20. $A = \frac{1}{2} \times 4.5 \times 6 = 13.5\ m^2.$

21. $A = \frac{1}{2}(40 + 60)(30) = \frac{1}{2}(100)(30) = 1{,}500\ m^2.$

22. $V = 20 \times 10 \times 12 = 2{,}400\ in^3.$

23. $|5 - (-7)| = |12| = 12\ units.$

24. $Base = 9 - 1 = 8.\ Height = 7 - 1 = 6.\ Area = \frac{1}{2} \times 8 \times 6 = 24$ square units.

25. $SA = 2(10)(8) + 2(10)(3) + 2(8)(3) = 160 + 60 + 48 = 268\ m^2.$

26. On a single day, the total is one number. However, if she collects data across many days, the number would vary and become statistical. As asked (for one day), it has a single answer.

27. Total: $2 + 5 + 7 + 4 + 1 = 19.$ The tallest bar is 20–29 minutes. The data is roughly symmetric around that interval, with a slight skew to the right.

28. A dot plot places one dot for each data value, so you can see every individual value.

29. Short whiskers and a narrow box indicate small range and small IQR, meaning data values are close together.

30. Group Q has a higher median $(60 > 45)$ and a smaller IQR $(5 < 20)$, so it is both higher and more consistent.

✓ Practice Test 8 — Answer Key

1. B

2. Part A: 0.5 laps per minute (or 1 lap every 2 minutes).   Part B: 7 laps.

3. 15

4. B

Find more at
ViewMath.com/CO-Grade6

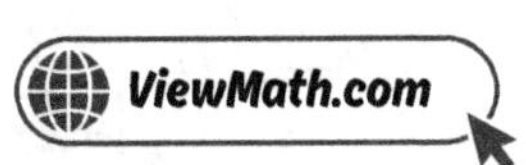

5 40 minutes   **6** C   **7** C   **8** D   **9** B   **10** -4 **11** B   **12** D

13 $(2+3) \times 4 = 20$ **14** $3n - 8$ **15** 25

16 The student forgot to distribute 4 to the 3. Correct: $8x + 12$ **17** $100 - 8h$ **18** C

19 Answers vary. Example: A classroom has at most 15 computers.   **20** A   **21** 27 m^2 **22** C

23 C   **24** 42 square units   **25** 6 cm   **26** C

27 Skewed to the left (or: most data on the right with a tail to the left).   **28** B

29 Range $= 55$, IQR $= 30$ **30** A

💡 Time to Learn! 💡

Review the explanations below, **especially for the questions you missed**.

Understanding why each answer is correct builds stronger problem-solving skills.

Tip: Circle any questions you got wrong, then read their explanation carefully.

📖 Practice Test 8 — Detailed Explanations

1 Dogs have 4 parts. Each part $= 2$ animals. Dogs $= 4 \times 2 = 8$.

2 Part A: From the graph, the swimmer completes 1 lap every 2 minutes, so the rate is $\dfrac{1}{2}$ lap per minute. Part B: $14 \div 2 = 7$ laps.

3 $2 \times 5 = 10$ strawberries, so $3 \times 5 = 15$ cups of yogurt.

4 x doubled from 4 to 8, so y doubles from 7 to 14.

Find more at
ViewMath.com/CO-Grade6

5 $1{,}000 \div 250 = 4$ batches. $10 \times 4 = 40$ minutes.

6 1 yard $= 3$ feet. $7 \times 3 = 21$ feet.

7 $I = 1{,}000 \times 0.03 \times 4 = \$120$.

8 Plan A: $\$60 \times 4 = \$240$. Plan B: $\$75 \times 4 = \$300$. Difference $= \$300 - \$240 = \$60$.

9 The GCF of 30 and 45 is 15. Divide: $30 \div 15 = 2$ and $45 \div 15 = 3$. So $30 + 45 = 15(2 + 3)$. Choice A uses 5, which is a common factor but not the greatest.

10 Going down from ground level means going below zero. Four floors down is -4.

11 City A shows $-20°$ and City B shows $-5°$. Since $-5 > -20$, City B is warmer. Think of the thermometer: the higher the mercury, the warmer the temperature. City A's mercury is lower, so it is colder.

12 Both vertices share $x = 1$, so the side length is $|3 - (-2)| = |3 + 2| = 5$. Choice B uses only $|3 - 0|$. Choice C computes $|3 - (-1)|$ or a similar error.

13 Without parentheses: $2 + 12 = 14$. With parentheses: $(2 + 3) \times 4 = 5 \times 4 = 20$.

14 Start with n. Multiply by 3: $3n$. Subtract 8: $3n - 8$.

15 $4^2 + 2(4) + 1 = 16 + 8 + 1 = 25$.

16 $4(2x + 3) = 4 \times 2x + 4 \times 3 = 8x + 12$, not $8x + 3$.

17 Start at 100 and subtract 8 per hour. $100 - 8h$.

Find more at
ViewMath.com/CO-Grade6

18 $3(5) = 15 \neq 18$. So $x = 5$ is not a solution. The correct solution is $x = 6$.

19 Any situation where a quantity is 15 or fewer works.

20 Triangle A: $\frac{1}{2} \times 4 \times 4 = 8$. Triangle B: $\frac{1}{2} \times 5 \times 3 = 7.5$. Triangle A has the greater area.

21 $A = 4.5 \times 6 = 27 \ m^2$.

22 Volume of a rectangular prism is $V = l \times w \times h$.

23 When two points have the same y-coordinate, the segment between them is horizontal.

24 Length $= |9 - 2| = 7$. Width $= |5 - (-1)| = 6$. Area $= 7 \times 6 = 42$ square units.

25 $6s^2 = 216$, so $s^2 = 36$, giving $s = 6 \ cm$.

26 Exercise time varies from student to student. The other questions each have one fixed answer.

27 The peak is at 5, with values trailing to the left toward 1. Most data is on the higher end.

28 With 200 data points, a dot plot would be too crowded. A histogram groups the data into intervals for a clearer display.

29 Range $= 80 - 25 = 55$. IQR $= 65 - 35 = 30$.

30 Class A's MAD is 3, meaning scores are typically within 3 of the mean. Class B's scores typically deviate 9 points. Class A's student is more likely to be near 75.

✅ Practice Test 9 — Answer Key

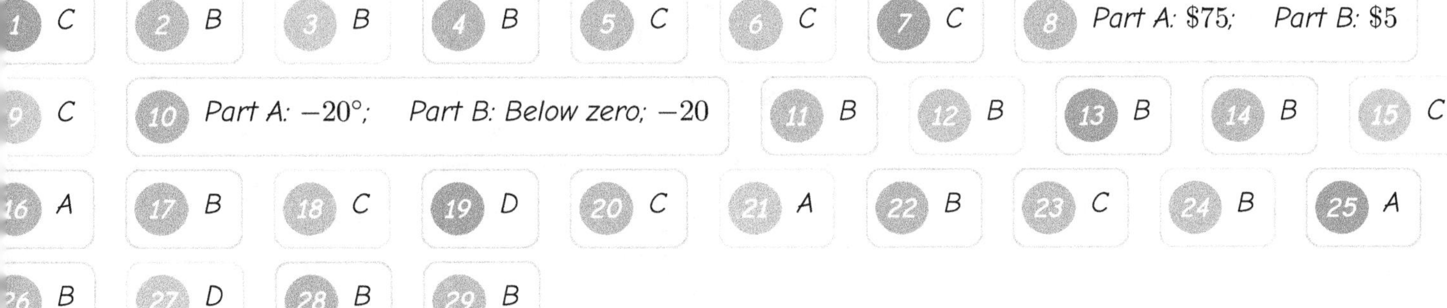

1 C	**2** B	**3** B	**4** B	**5** C	**6** C	**7** C	**8** Part A: $75; Part B: $5

9 C **10** Part A: $-20°$; Part B: Below zero; -20 **11** B **12** B **13** B **14** B **15** C

16 A **17** B **18** C **19** D **20** C **21** A **22** B **23** C **24** B **25** A

26 B **27** D **28** B **29** B

30 Both have mean = 22. Player A range = 8; Player B range = 25. Same average, but Player A is more consistent.

💡 Time to Learn! 💡

Review the explanations below, **especially for the questions you missed**.

Understanding why each answer is correct builds stronger problem-solving skills.

Tip: Circle any questions you got wrong, then read their explanation carefully.

📖 Practice Test 9 — Detailed Explanations

1 Total parts $= 1 + 3 + 2 = 6$. Each part $= 18 \div 6 = 3$. Blue $= 3 \times 3 = 9$.

2 Divide: $120 \div 4 = 30$ pages per minute.

3 $3 \times 2 = 6$, so $5 \times 2 = 10$.

4 $1 : 3$ multiplied by 3 gives $3 : 9$, so $(3, 9)$ is on the line.

Find more at
ViewMath.com/CO-Grade6

5 $3 \times 4 = 12$ *raisins, so* $4 \times 4 = 16$ *cups of granola.*

6 $1\ L = 1,000\ mL.\ 2.5 \times 1,000 = 2,500\ mL.$

7 *Interest* $= \$500 \times 0.18 = \$90.$ *New balance* $= \$500 + \$90 = \$590.$

8 *Part A: Friend A* $= \$40 \times 5 = \$200.$ *Friend B* $= \$25 \times 5 = \$125.$ *Difference* $= \$200 - \$125 = \$75.$ *Part B:* $I = 125 \times 0.04 \times 1 = \$5.$

9 $4(9 + 6) = 4 \times 9 + 4 \times 6 = 36 + 24 = 60.$ *You can also compute* $4 \times 15 = 60.$

10 *The mercury reaches the* $-20°$ *mark, which is below the* $0°$ *line. Below zero is negative, so the temperature is* $-20.$

11 *A higher temperature reading means warmer. Since* $-2 > -5$ *(think of a thermometer.* $-2°$ *is above* $-5°$*), Tuesday was warmer. Monday's* $-5°F$ *is colder because it is farther below zero.*

12 *The two points share the same* y-*coordinate, so the distance is the absolute difference of the* x-*coordinates:* $|3 - 7| = |-4| = 4.$ *Choice A subtracts without taking the absolute value. Choice C adds the* x-*coordinates.*

13 *Parentheses:* $4 + 4 = 8.$ *Divide:* $48 \div 8 = 6.$

14 *"7 less than* y*" means subtract 7 from* y*:* $y - 7.$ *The order is reversed from how it reads.*

15 $3^2 + 4^2 = 9 + 16 = 25.$

16 $9w - 4w = 5w$ *and* $6 - 6 = 0.$ *Result:* $5w.$

17 *Notebooks cost* $3n$ *and pens cost* $1 \times p = p.$ *Total:* $3n + p.$

Find more at
ViewMath.com/CO-Grade6

18 Multiply both sides by 4: $m = 7 \times 4 = 28$.

19 "No more than 200" means 200 or fewer: $c \leq 200$.

20 $A = \frac{1}{2} \times \frac{1}{2} \times \frac{1}{4} = \frac{1}{16} \; ft^2$.

21 Rectangle: $8 \times 5 = 40$. Triangle: $\frac{1}{2} \times 8 \times 3 = 12$. Total: $40 + 12 = 52 \; cm^2$.

22 $V = 5 \times 3 \times 4 = 60 \; cm^3$.

23 Same y-coordinate. $|4 - (-2)| = 6$ units.

24 Base $= 8$, height $= 6$. Area $= \frac{1}{2} \times 8 \times 6 = 24$ square units.

25 $2(4 \times 3) + 2(4 \times 2) + 2(3 \times 2) = 24 + 16 + 12 = 52 \; cm^2$.

26 Students live at different distances, so travel times vary. This is what makes it a statistical question.

27 Most values are between 2 and 7, clustered around 4. The value 20 is far away from the rest, making it an outlier.

28 The height of a histogram bar tells you how many data values fall in that interval. A height of 8 means 8 values.

29 10 values. Median $= (9 + 11) \div 2 = 10$. Lower half: $1, 3, 5, 7, 9$. $Q1 = 5$. Upper half: $11, 13, 15, 17, 19$. $Q3 = 15$. $IQR = 15 - 5 = 10$.

30 Player A: $(18 + 20 + 22 + 24 + 26) \div 5 = 22$, range $= 8$. Player B: $(10 + 15 + 22 + 28 + 35) \div 5 = 22$, range $= 25$. Both average 22 points, but Player A's scores are much more consistent.

Find more at
ViewMath.com/CO-Grade6

✅ Practice Test 10 — Answer Key

1 C   **2** B   **3** B   **4** A   **5** C

6 Part A: 2.8 m;   Part B: 2,800 mm;   Part C: Longer by 0.8 m (80 cm).   **7** $1,260   **8** 6 weeks

9 B   **10** B

11 Part A: $J = -3.5$, $K = 1.5$, $L = -1$, $M = 0.5$, $N = -2.5$. Part B: J, N, L, M, K. Part C: J has the greatest a

12 C   **13** $5^2 = 25$ square units   **14** C   **15** B   **16** $12x - 5$ **17** D   **18** $3x = 18$; $x = 6$

19 C   **20** B   **21** B   **22** B   **23** $(9, -5)$ **24** A   **25** B   **26** Statistical   **27** C

28 C   **29** A   **30** B

💡 Time to Learn! 💡

*Review the explanations below, **especially for the questions you missed**.*

Understanding why each answer is correct builds stronger problem-solving skills.

***Tip:** Circle any questions you got wrong, then read their explanation carefully.*

📖 Practice Test 10 — Detailed Explanations

1 Apple has 3 parts, orange has 5 parts. Total parts = 8. Total ounces = $8 \times 4 = 32$.

2 Store A: $4.50 \div 10 = 0.45. Store B: $3.20 \div 8 = 0.40. Store B is cheaper per pencil.

Find more at
ViewMath.com/CO-Grade6

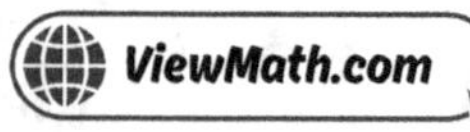

3 $2:3$ multiplied by 3 gives $6:9$. The other pairs do not simplify to $2:3$.

4   The ratio is $2:3$. Choice A continues with $6:9=2:3$. Choice B has $6:8$ which is not $2:3$.

5 $42\div 6=7$ sections, which needs $7+1=8$ posts (one at each end). Note: if counting "per 6 feet" as a rate, $42\div 6=7$, but fences need a post at the start too, giving 8.

6   Part A: $280\div 100=2.8$ m. Part B: $280\times 10=2{,}800$ mm. Part C: $2.8-2=0.8$ m longer.

7   Monthly savings: $0.30\times \$350=\$105$. Yearly savings: $\$105\times 12=\$1{,}260$.

8   Weekly savings $=0.50\times \$40=\$20$. Number of weeks $=\$120\div \$20=6$ weeks.

9   Factors of 8: $1,2,4,8$. Factors of 36: $1,2,3,4,6,9,12,18,36$. The greatest common factor is 4. So $36+8=4(9+2)$.

10   On a horizontal number line, negative numbers are always to the left of zero and positive numbers are to the right.

11   Part A: Read each point from the number line using the half-unit marks. Part B: Order the values: $-3.5 < -2.5 < -1 < 0.5 < 1.5$, so J, N, L, M, K. Part C: Find absolute values: $|J|=3.5$, $|K|=1.5$, $|L|=1$, $|M|=0.5$, $|N|=2.5$. The greatest is 3.5, belonging to Point J. Note that J is the least value but has the greatest absolute value — absolute value and number order are different!

12   The lights share $y=5$, so the distance is $|-3-4|=|-7|=7$ blocks. Choice B uses only the y-coordinate. Choice D adds $|-3|+|4|+2$ or a similar error.

13   The square has side length 5, so area $=5^2=25$ square units.

14   Half of n is $\frac{n}{2}$. "6 more" means add 6: $\frac{n}{2}+6$.

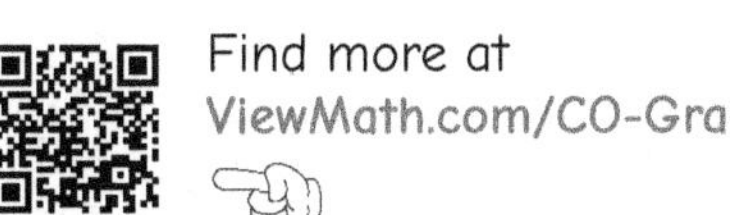

15. $10 - 2(4 - 1) = 10 - 2(3) = 10 - 6 = 4.$

16. $9x + 3x = 12x$ and $2 - 7 = -5$. Result: $12x - 5$.

17. Splitting \$14 equally among f friends means dividing: $14 \div f$.

18. Left: 3 circles $= 3x$. Right: 18 unit squares $= 18$. Equation: $3x = 18$. Divide: $x = 6$.

19. $n \leq 8$ means n is 8 or less. $8 \leq 8$ is true. 8.5, 9, and 10 are all greater than 8.

20. $24 = \frac{1}{2} \times 8 \times h$, so $24 = 4h$, which gives $h = 6$ ft.

21. $A = b \times h = 9 \times 5 = 45$ cm^2.

22. $V = 3 \times 2 \times 2 = 12$ ft^3.

23. The fourth vertex must share an x-value with $(9, 3)$ and a y-value with $(2, -5)$, giving $(9, -5)$.

24. Bottom rectangle: $8 \times 3 = 24$. Top-left rectangle: $4 \times (7 - 3) = 4 \times 4 = 16$. Total: $24 + 16 = 40$ square units.

25. $SA = 2(20)(30) + 2(20)(8) + 2(30)(8) = 1200 + 320 + 480 = 2{,}000$ cm^2.

26. Different students send different numbers of texts, so the answers vary.

27. An outlier is a value that is far from the rest of the data set.

28. 3 hours: 5 dots, 4 hours: 3 dots, 5 hours: 2 dots. Total $= 5 + 3 + 2 = 10$ students.

Find more at
ViewMath.com/CO-Grade6

29. *Min = 8. Q1: median of 8, 12, 14 = 12. Median = 18. Q3: median of 22, 26, 30 = 26. Max = 30.*

30. *A good summary combines center ("typical value is about 20") and spread ("span 10 units").*

Well done checking your answers!

Keep practicing to strengthen your skills.

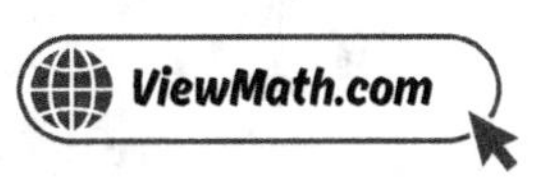

Author's Final Note

I hope you enjoyed this book as much as I enjoyed writing it. Whether you are a student working through the materi
a parent supporting your child's learning, or a teacher guiding your class, I have tried to make this book as clear and
engaging as possible. I hope I have succeeded. If you have any suggestions for improvement, please let me know. I
would love to hear from you.

The accuracy of calculations is very important to me. We have done our best, but I also expect that I have made som
minor errors. Constant improvement is the name of the game. If you find any errors, please let me know. I will fix
them in the next edition.

For students: Your learning journey does not end here. I have written a series of books to help you learn math. Mak
sure you browse through them. I especially recommend workbooks and practice tests to help you prepare for your
exams.

For parents: Thank you for investing in your child's education. I encourage you to explore the companion resources
available online to help support your child outside the classroom.

For teachers: Thank you for the invaluable work you do every day. I hope this book serves as a useful resource in yo
classroom. Feel free to reach out if you have suggestions or would like to discuss how best to use this book with your
students.

I also enjoy reading your reviews. If you have a moment, please leave a review on where you found this book. It will
help others find this book. If you have any questions or comments, please feel free to contact me at
drNazari@ViewMath.com.

And one last thing: Remember to use online resources for additional help. I recommend using the resources on
https://ViewMath.com You can find video lessons, practice problems, and more. You can also use the online
companion for this book to track your progress and access additional resources.

Wishing all students the best in their studies, parents every success in supporting their children, and teachers
continued inspiration in their classrooms!

Dr. A. Nazari

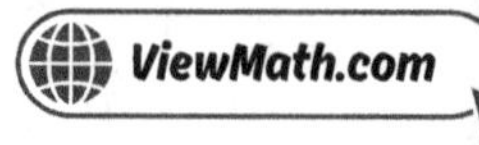

Great Job! Keep Learning with ViewMath!

Keep up the great work! Visit **viewmath.com/CO-Grade6** for free lessons, quizzes, and more.

Study Guide

Workbook

Step-by-Step

3 Practice Tests

5 Practice Tests

7 Practice Tests

Find more at
ViewMath.com/CO-Grade6

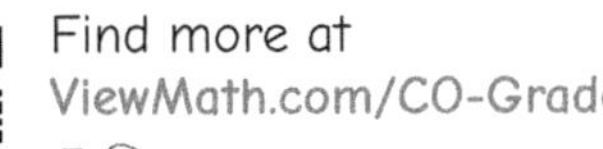

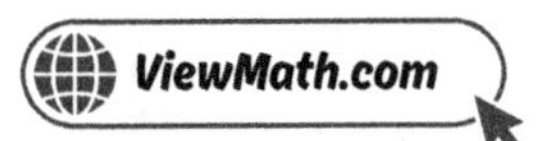